Help for the Technology Shy
presents...

Your Technology Path to Follow

HOW TO GET STUFF DONE WITH YOUR COMPUTER, TABLET OR SMARTPHONE

Michael Gorzka (Writer)
Tanya Zeinalova (Illustrator)

"LET'S HIT THE ROAD"

www.technologyshy.com

International Standard Book Number:
ISBN: 978-1-7349815-7-5

The author and publisher advises readers to take full responsibility for their technology use and have neither liability nor responsibility to any person or entity with respect to any loss or damages arising from the information contained in this book.

Table of Contents

Preface: Soap Boxing, Painting in Broad Strokes and Use of Icons & Italics

"SOAP BOX MEETING"

"Therefore, since brevity is the soul of wit

And tediousness the limbs and outward flourishes,

I will be brief..." — Polonius

Soapboxing (hopefully not too much)

"SOAP BOX, DIFFERENT SOCKS"

The information and conclusions in this book are based on my 25+ years experience teaching adults how to use (& not misuse) various and sundry devices, applications & online services. Please excuse my occasional pontifications ☺️

Note about the illustrations

Many of Tanya's original illustrations for this book are captioned with the fun file names she gave them.

"Me"

Due to space considerations, the examples in this book will be painted in "broad strokes" (so to speak).

Supplemental blog posts and videos for this book (to fill in the gaps as it were) are available on the Help for the Technology website.

Italicized words and phrases are specific Help for the Technology Shy blog / video topics (see the "Where to go from here" chapter for details).

"THE BLOG, WIZARD'S MUG"

Notes From the Help for the Technology Shy "Tough Love" Department

"TOUGH LOVE"

If you are struggling with contemporary technologies, rest assured you are in **very** good company, my friend❣️

Tech Wizard Mike (c'est moi) has worked with (or frankly sometimes **tried** to work with) some very intelligent people including college professors, community activists 🕊️, well-known authors, fellow librarians (🤫) and even a NASA engineer (🛰️) who all struggled mightily with twenty-first century technologies or were simply unable to make productive use of them due to...

- lack of clear objectives
- lack of focus
- unrealistic expectations
- willy-nilly, full-title boogie, college party animal, full-steam potato use of the technology basics 🐭 ⌨️ 👆
- not applying real world commonsense and methodologies to their technology use
- a belief that computers, tablets, smartphones et al. are smarter than their human users (which they are most definitely **not**)
- using out-of-date (i.e. *obsolete*) devices
- falling for scams 😈
- not taking personal responsibility for their own devices
- losing sight of the *forest through the trees*
- lack of acceptance (some technologies are simply poorly designed, kludgy and/or buggy)
- And, lest we forget, good old-fashioned *obstinance*

These smart cookies did not apply what we have learned about how to do things in **real life** to their technology-related endeavors.

It can be reasonably assumed that when any of the aforementioned clever folks make a salad 🥗, ride a bicycle 🚴, build a structure 🧱, cross an intersection 🚦, knit a sweater 🧶 or make a cup of tea 🫖 ...

They would do so with a clear objective / goal / target in mind.

They would proceed in a logical order (e.g. they get on the bicycle **before** trying to ride the bicycle).

They would not be in an arse-busting rush ❗

"Haste makes waste" and "slow and steady win the race" are **exponentially true** when it comes to contemporary technology use 🐢 🏁

These astute individuals would have **realistic expectations** throughout (e.g. they would not expect their various salad components to fall from the sky into their salad bowls 🥒).

And finally, they would not want to depend on *step-by-step instructions* for such things as crossing a busy intersection which is actually a **pretty complex operation** that we routinely accomplish by following visual cues with a clear objective while using caution and commonsense.

“CROSSWALK”

If the preceding list makes at least some sense to you 🤔, then you're golden, you're good to go, you're off to the races 🎠, you are exceptionally mentally and emotionally prepared to use twenty-first century technologies 👍

If you don't feel you **are** exceptionally mentally and emotionally prepared to use twenty-first century technologies then read on 😊

"WORRIED DUCK"

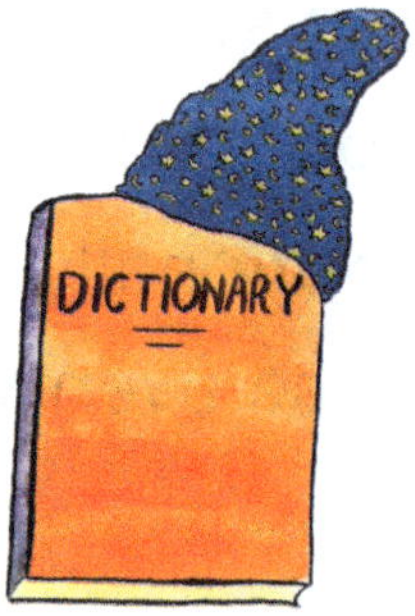

dip your toe in the water

to start very carefully to do or become involved in something that you are not experienced at.

Presenting Your Very Broad Outline for Productive, Stress-Free Technology Use

"Pouring a cup of tea"

"Blanket Nirvana"

* Brew a nice cup of tea ☕

* Remember to *breathe.*

* Master the easy peasy lemon squeezy *timeless technology basics* — while keeping in mind that they are powerful tools that must be used with thought, intention and focus 🤔 🎯

* Know *which device you are using* **and** *which operating system* (or "OS") is running on that device.

* Noodle around with any **devices** that you need to use in order to get familiar with them 🖥️ 💻 📱

* Noodle around with any **apps** that you need to use in order to get familiar with them

* Visualize and sketch out your targets, your goals, your objectives, your visualized outcomes 🙌 (however you wish to think of them).

* Understand that we need to use our twenty-first devices and online services with the same mindset we use to brew a nice cup of tea 🫖

* Think of technology as a whole (in the form of devices, apps and online services) as simply a means to reach your desired goals 🥅

* Realize that nothing is engraved in stone 🗿 and it pays dividends to be flexible and peaceful ✌️

* You understand *conceptually* the task at hand **before** you reach for any of your mobile devices or go anywhere near a computer 🙇

* You use "breadcrumbs" and other visual cues to navigate devices, apps and websites 🍞👀

* You *document your technology use* 📙✏️

* You stay focused on the task at hand while using any device, app or online service 🎯

* You see the *forest through the trees* 🌲

* You expect the unexpected, accept life's imperfections and go with the flow 🏞️ 🛶

* You get *good help* if you need it 🤨 → 🔎

* You avoid the *Sisyphus syndrome* 📋

* And you approach any technology-related challenge gradually, **step-by-step** while being prepared to (figuratively of course) fall on your face a few times:

"CROW FAIL"

Breathe.

"BREATHE + ELEMENTS"

The Breath/Health Connection

> Every system in the body relies on oxygen. From cognition to digestion, effective breathing can not only provide you with a greater sense of mental clarity, it can also help you sleep better, digest food more efficiently, improve your body's immune response, and reduce stress levels. Many people who practice yoga, especially those who focus on the practice of Pranayama, or breath control, have been shown to regulate the sympathetic nervous system, or the part of the body that controls breathing, heart rate and blood pressure.[1]

We need to apply real world commonsense, problem-solving and **breathing** to our contemporary technology use.

You will have reached the point of diminishing returns the moment you begin to feel more than moderate *stress*; the moment you begin to "white knuckle" the device you are using.

---

[1] "The Importance of Proper Breathing for Your Overall Health." Elliott Physical Therapy, https://elliottphysicaltherapy.com/importance-proper-breathing-overall-health/. Accessed 28 September 2020.

The moment you see your knuckles turn the slightest shade of white, please **take a break** — all of the twenty-first century technologies will still be here when you get back.

We should note here that being calm, cool and collected during technology use is not an option. **It is a necessity**.

Visit your local library for resources on how to stay calm and regain composure.

How to Make the Perfect Cup of Tea 🫖

"ART OF BREW"

1. Treat your water kindly. Run the tap a little so the water's nicely aerated, and only boil it once to keep the oxygen level up...
2. Add tea and water. Pop a tea bag into your mug, pour over the hot water and stir briefly.
3. Wait patiently...
4. Give the tea bag a squeeze...
5. Customize your brew with milk and/or sugar.
6. Enjoy.[2]

---

[2] "How to make a proper brew." Taylors of Harrogate Yorkshire Tea, https://www.yorkshiretea.com/our-teas/how-to-make-a-proper-brew. Accessed 9/26/20. Accessed 26 September 2020.

Question: What does proper tea brewing have to do with productive, stress-free technology use?

Answer: Everything, really.

When we brew a cup of tea, we have a *visualized outcome* in mind 🙌

We proceed *logically* and *methodically*. We would not, for example, pour the hot water **before** heating it.

We would wait *patiently* for the tea to steep before drinking it.

We would also (again, thinking logically) add the milk and sugar **after** removing the tea bag from the cup.

"There is something in the nature of tea that leads us into a world of quiet contemplation of life." – Lin Yutang

It also pays to be *flexible.*

What if, for example, there are no more tea bags on the premises but we have just discovered a tin of loose tea…

And (because we are **observant** as well as **methodical** while we are in tea brewing mode) we see we are well-within the "best by date" on the tin.

We need to approach computers, tablets, smartphones, wearables and any other high-tech gadgetry with the same mindset — because, if the truth be known, those devices are all as dumb as a box of rocks or a bag of hammers (pick your analogy).

dumb as a box of rocks

adjective

extremely unintelligent.

"TABBY WITH BOX OF STONES"

Get a Technology Notebook

"TABBY THE BOOK SPREADER"

A cleverly disguised (🤫) passwords book / technology journal such as "Tabby's Guide to Thoughtful Tea Drinking" will help you keep track of your various and sundry usernames, passwords / passphrases, security questions and so forth.

And it will also give you a place to record any technology-related questions, distractions, conundrums and kerfuffles that you come across.

Get things to remember, things to keep track of, and things that distract you out of your head and into your *technology notebook* — so that you can resume getting stuff done with whichever computer / mobile device / application / online service you were using.

→ Record your device's "About" info into your technology notebook.

→ Use it for **reference** when you are researching issues with your device or how to carry out a particular task with your device.

→ Review your passwords book / technology journal on a daily basis.

→ Make it your bedside reading until you feel 💯% comfortable with your technology use.

→ And after that, review your technology journal periodically.

→ And, of course, use it to record your favorite tea-drinking experiences (or Tabby will be very disappointed 😿).

"Documentation is a love letter that you write to your future self."

Your Technology Foundation[3]

"BRICK TEAM"

1. Know the *geeky terms* (e.g. "downloading" vs. "streaming").
2. Master the *timeless technology basics.*
3. Find out and document *what you are using*.
4. Make sure your device is free from *malware*.
5. Have a system in place to keep track of your *passwords*.

---

[3] Michael Gorzka, Help for the Technology Shy,(Newton: Squirrels Drinking Coffee, 2020), 98–104.

"Timeless" Technology Basics Review

We call these the "timeless" technology basics because, unlike many other things related to technology, they have remained relatively **constant** over the years.

"BALL"

While we're on the subject of constants, we should add that *cause & effect* and *GIGO* have always been — and always will be — major components of technology use.

If you do not feel comfortable with any of these technology basics, an instructional DVD is available 💿 (Please see the "Where to go from here" chapter for details).

Without further adieu, the timeless technology basics are... 🥁

1. point and click the mouse 🐭 (or trackpad)
2. tap the screen, swipe the screen, pinch the screen (ouch! 😟)
3. press keys on the keyboard

The most important takeaway from this chapter is that **actions** (in the form of the technology basics) will often have **reactions**.

Our physical actions in the real world will often yield some type of **result** — which in general will be positive if carried out with thought & intention and negative when performed in a thoughtless and irresponsible manner❣️

When you click the mouse, tap a screen or press keys on a keyboard, you are sending your device a signal, you are providing **input** (of some kind).

COMPUTER "MICE" WILL VARY IN APPEARANCE BUT THEY ARE ALL INPUT DEVICES

KEYBOARDS OFFER SPECIAL FUNCTION KEYS AND KEYBOARD SHORTCUTS

WE HAVE MANY OPTIONS! (E.G. YOU MAY FEEL MORE COMFORTABLE USING A STYLUS THAN YOUR FINGER)

Context Matters! (For example, this young lady is not in her Math class)

Powerful tools

"TOOLS FOR STUFF DONE"

The technology basics are *easy peasy lemon squeezy* to learn and to carry out — but they are also very **powerful tools** which must be used with thought and intention while always being well-aware of your *context*.

Your Tech Wizard Scribe has worked with **many** people over the years (since 1995) who had used the *timeless technology basics* to click or tap or keyboard their way into confusion, lost work, wasted time, malware infestation and even *identity theft*.

Examples of *context* are...

- which app is currently *active*
- which folder on your device is currently open
- what the tip of your screen arrow or finger is currently over
- where you are within your device or online service

The following examples are from macOS, Windows 10 and iOS.

Please keep in mind regardless of which device (or online service) you are using, always look for **visual cues** to see where you are, because *context* can effect the **outcome** of any mouse click, screen tap or keyboard press.

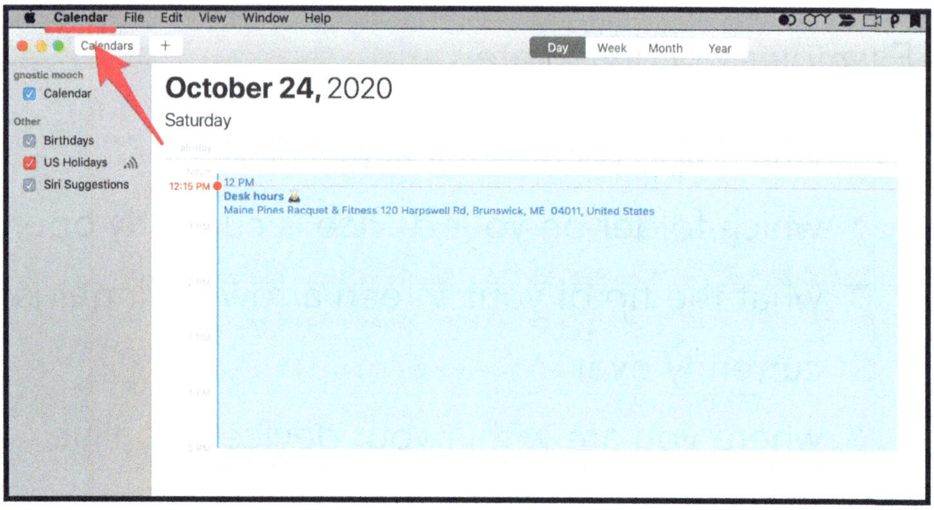

Calendar is currently the "active" application on this macOS computer.

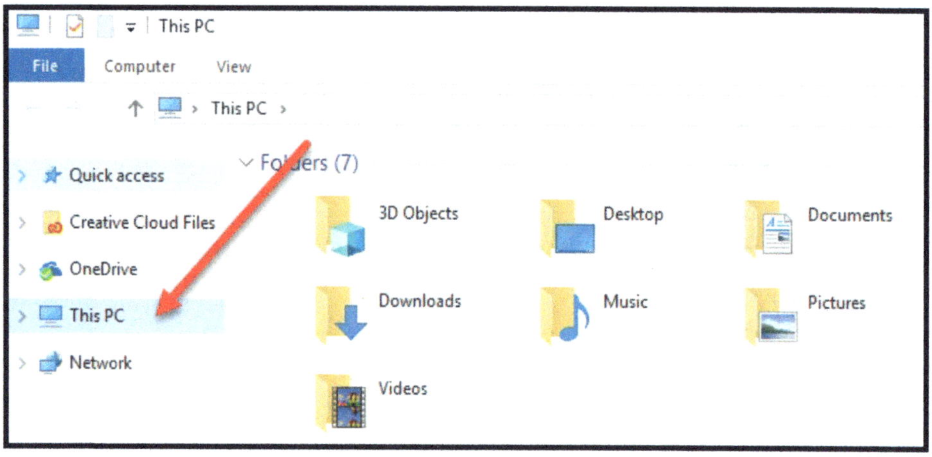

USE VISUAL CUES TO GET A SENSE OF YOUR DEVICE'S FILE SYSTEM

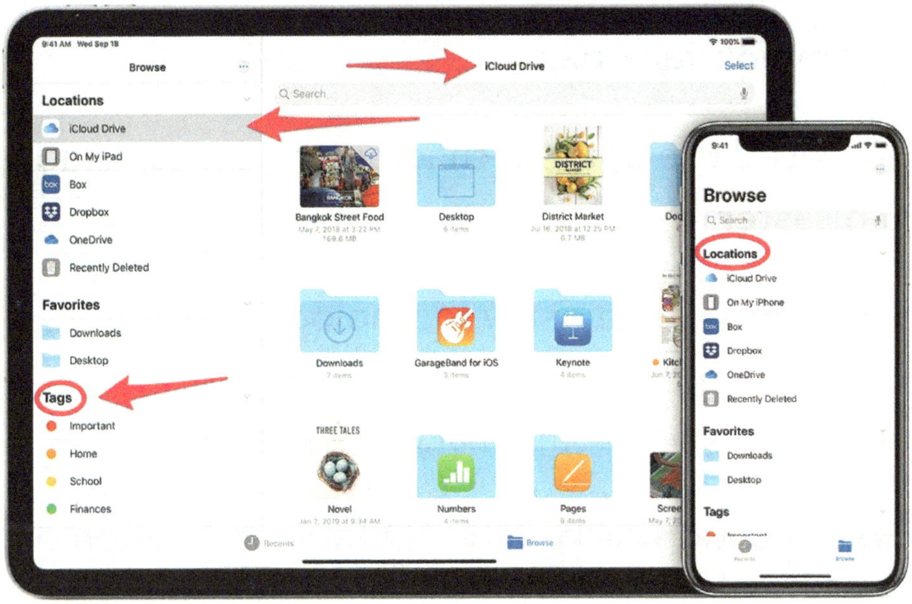

Locations, files, folders & functionality (e.g. "Tags") — it pays to be visual 👀

All roads lead to rome.

JUST LIKE IN REAL LIFE, VISUAL CUES WILL GUIDE YOU DURING TECHNOLOGY USE

A common faux pas

Ill-advised use of the technology basics can lead to **disaster** 🙀

For example, one of my technology-shyest friends fell for the mother of all (apparently legal 😒) flimflams aimed at the macOS platform.

First, Brandon clicked the "Download" button on a web page that looked something like this…

"MACGUARDIAN"

Don't do that!

And **then** Brandon double-clicked this installation icon which was downloaded from that unethical website:

MacGuardian.5.3.pkg

And as a result, turned his très expensive notebook computer into a high-tech paperweight.

Brandon was led astray be a professional-looking website and enticing promises. He should have stayed in his own "sandbox" (so to speak) namely his computer's *App Store*.

Just Noodling Around...

Knowing what you are using is one of your *technology foundation* building blocks.

"NOODLING AROUND, WATCHING OUT"

Do you have to use a particular device for your job?

Did you get your device as a gift? Did you buy it new? Did somebody give you their old device?

Do you use a particular public computer on a regular basis?

Regardless of how you came by it, it is important to have some general info about the device you are using so that...

- You can get *good help.*
- You can find out which peripherals and apps your device supports [4]
- You can determine if the device is *obsolete.*

Obsolete devices are going to be **vulnerable** to viruses and other forms of malware because they will no longer be receiving security updates.

And also, before you can become proficient with a computer / tablet / smartphone / wearable mystery gadget / whatsis, et al. you have to be **comfortable** with it.

And to be comfortable with a computer / tablet / smartphone / mystery gadget / whatsis, et al. you have to be **familiar** with it.

And to be familiar with any device, you need to know what **it** is.

---

[4] Gorzka, Michael. "Which peripherals and apps can you use with your device?." Help for the Technology Shy, www.technologyshy.com/whichapps. Accessed 2 October 2020.

And the first step to finding out what something **is**, is to casually say "hiya" 👋 to it.

Even if you already "know" your device (and are perhaps at odds with it 📱🥊), please follow this procedure anyway as the beginning is not only a great place to start, it's a great place to **regroup** (with a nice cup of tea ☕).

Simply brew that nice cup of tea and then use the click 🐭 or tap 👆 technology basics to thoughtfully and observantly browse around a device → simply to see what is there.

You are simply a passive observer during this "noodling around" phase. (No pressure 🙂)

You are not deleting, moving, changing or installing **anything**.

And although your are not being a secret squirrel about this, you **are** gathering nuts of information; filing things of interest, by perhaps, writing them down into your technology notebook. [5]

Searching for "About"

DOCUMENT ALL OF YOUR DEVICES AND PERIPHERALS ✏️

[5] Michael Gorzka, Tabby's Guide to Thoughtful Tea Drinking,(Akron: Squirrels Drinking Coffee, 2020).

You can search for your device's "About" info directly...

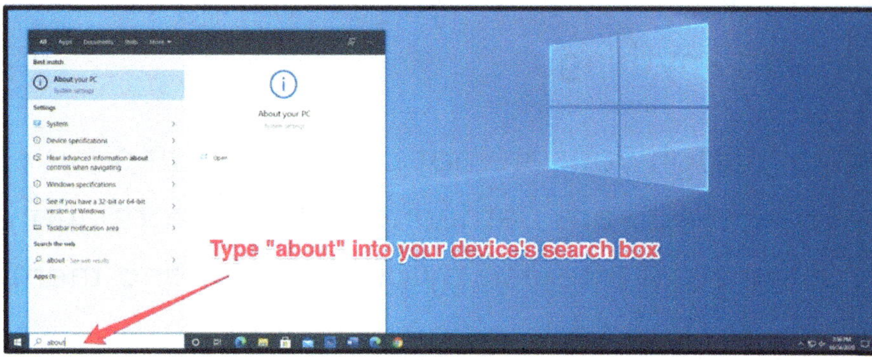

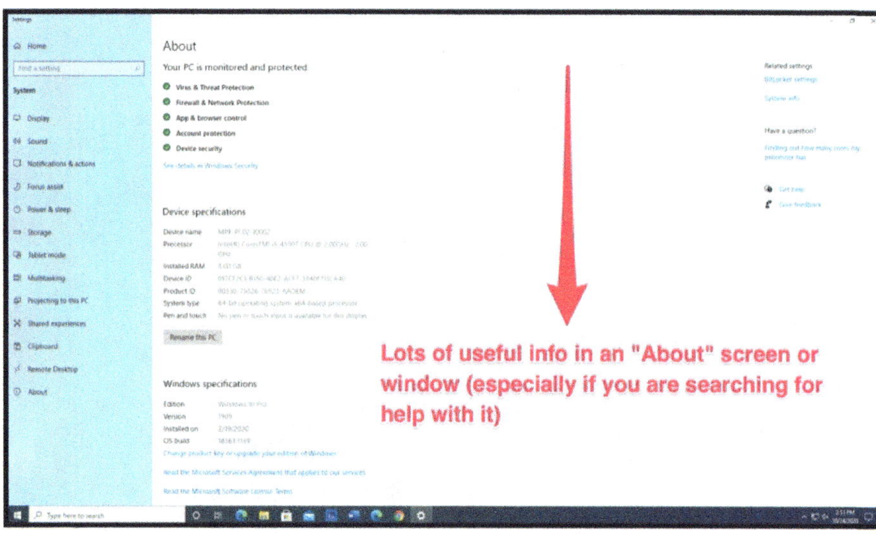

The location of a tablet's or a smartphone's *search box* can be a moving target so to speak. Methods to summon the search box will vary between devices and can actually change with software updates.

You can also casually *noodle around* your device until you see the magic word "About..." or at least a screen with info like this...

WRITE THIS INFO DOWN IN YOUR TECHNOLOGY NOTEBOOK. TAKE SPECIAL NOTE OF YOUR DEVICE'S CURRENT OS OR "SOFTWARE VERSION"

"Not all those who wander are lost."

—J. R. R. Tolkien"

What app?

You can perform this same "noodling around" procedure on any application — and you most definitely should noodle around any app that you need to use for your job.

You can browse through an app's menus to uncover "About" info and **useful functions**:

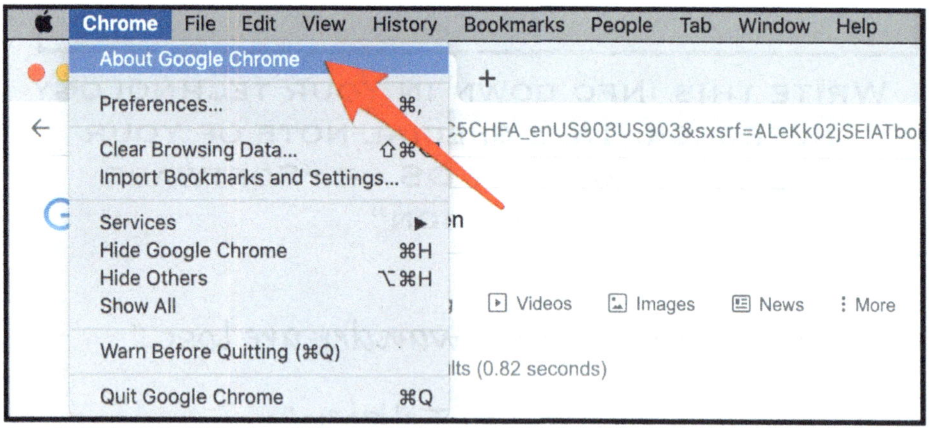

What's on the (Context) Menu?

"CHEF KNOWS BEST"

Context menu demos await you at www.technologyshy.com/context

Visual cues often reveal essential and indispensable navigation elements, context and functionality.

For example...

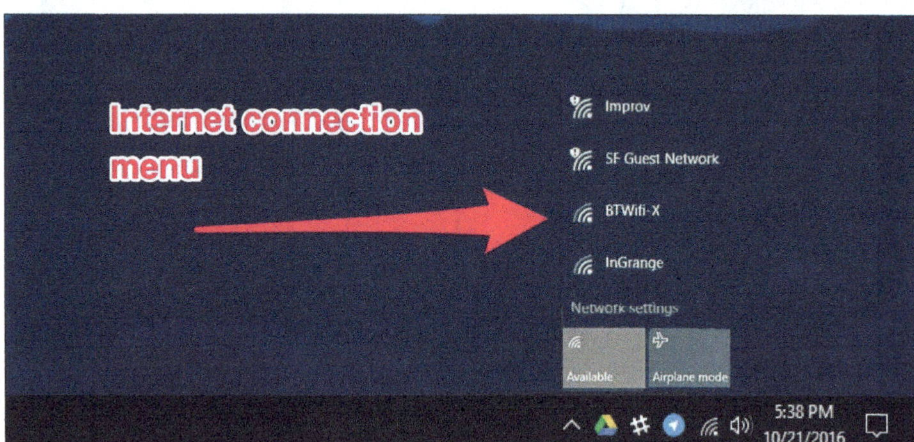

But what if there is something you want to do and there are no visual cues ❓🤷‍♀️

In that case, you could try a "secondary click" or a "tap & hold" (depending on your device) to summon a *context menu*:

DELETING AN APP FROM A TABLET WITH A CONTEXT MENU

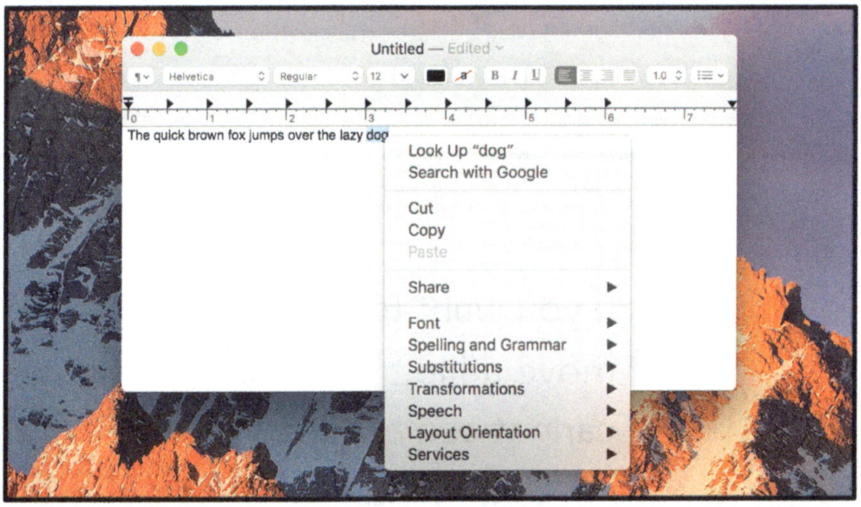

LOOKING UP A WORD WITH A CONTEXT MENU

Context matters (always)

The **selections** on a context menu will depend on and what the tip of a screen arrow (or finger on a touchscreen device) is currently over:

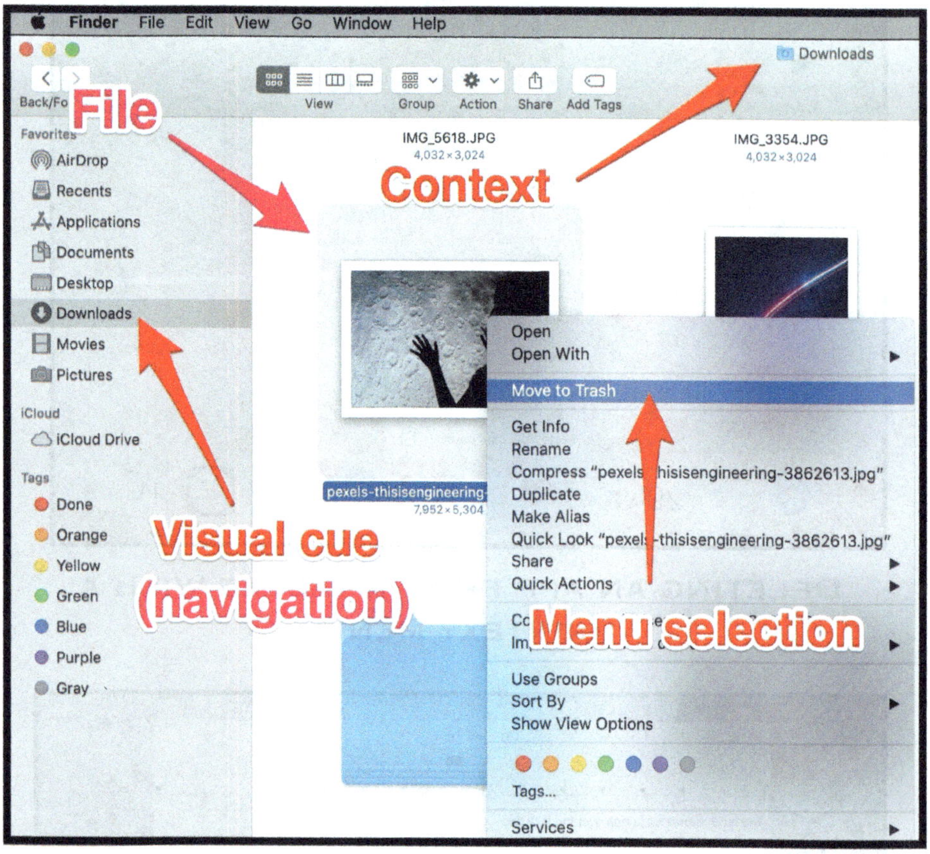

If you want to do something (e.g. move a file to the trash 🗑️) but cannot see a way to do it → try a context menu 🧑‍🍳

How to summon a context menu on your device?

1. You can gently *press your finger down* on a touchscreen device (e.g. a smartphone or a tablet) to summon a context menu.

2. If your computer or mobile device supports a mouse, you can click the right mouse button to summon a *context menu.*

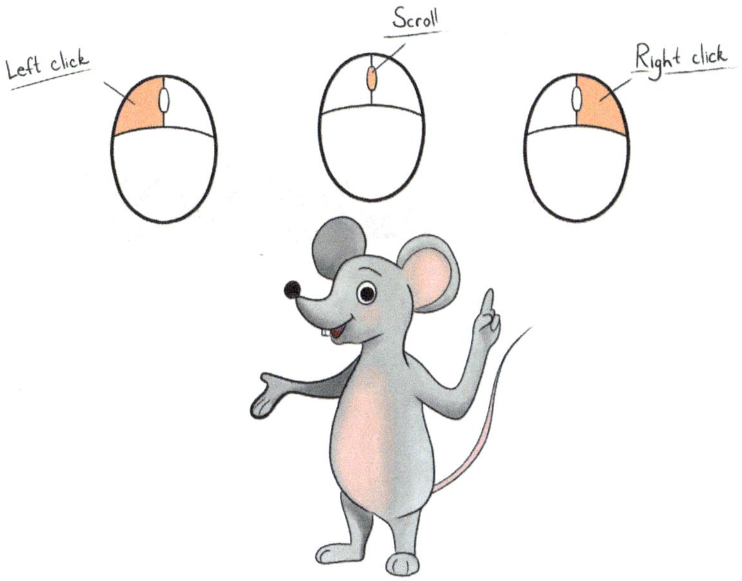

Look **carefully** at every piece of *hardware* (i.e. anything you can **physically touch**); analyze it; press buttons and see what they do in different contexts; look for *functionality* → such as the mouse scroll wheel and the dual buttons in the above illustration.

If your mouse does not have left & right mouse buttons, you may (depending on your device), be able to hold down the *control key* on the keyboard as you click the mouse or tap the trackpad:

YOU CAN SEARCH THE WEB FOR INFO ON YOUR DEVICE.

"Settings" might also be an option (depending on which computer or mobile device you are using):

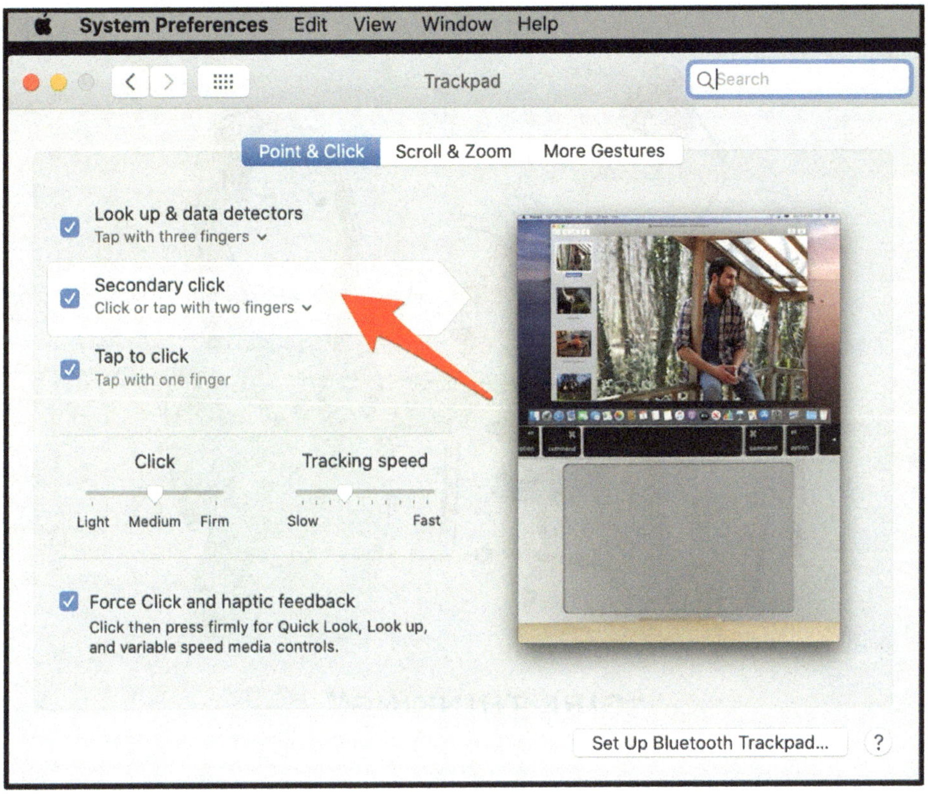

You may have come across your device's "Settings" or "System Preferences" or "Control Panel" as you were noodling around.

Objectives, Goals, Targets & Visualized Outcomes

"GIRL THINKING"

In other words, what would you like to do?

This is very important to know because technology proficiency begins with your defined goal 🥅 / target 🎯 / visualized outcome 🙌

There is no way to sugarcoat this statement from the Help for the Technology Shy "Tough Love" desk, so here is a treat in advance...

"C IS FOR COOKIE, L IS FOR LOVE"

Many people have disappointing, discouraging and downright disastrous experiences with their various gadgets and online services because they turn their brains off before using them.

This is no way to run a railroad because all contemporary technologies must be used mindfully; with thought and intention.

Patience is a virtue and a necessity. Keep your current written / sketched / doodled technology-related goal close by as you are **methodically** using the technology basics to work your way toward reaching that goal.

Simply, refer to it if any distractions arise.

You can record any technology-related distractions into your technology notebook (to be reviewed at a later date) thus enabling you to continue making progress toward your goal.

Seeing the Forest Through the Trees

"Hold on, where's the forest again?"

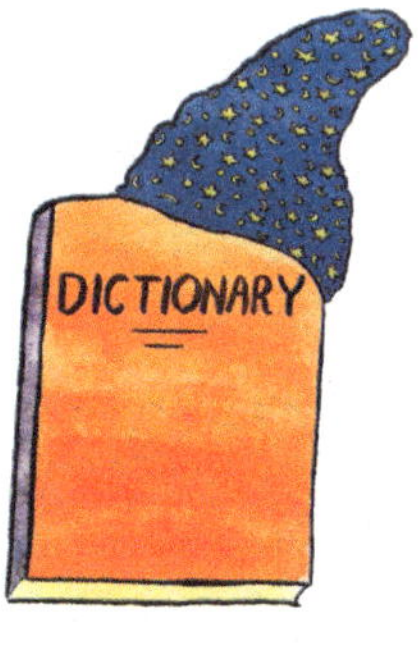

can't see the forest for the trees

An expression used of someone who is too involved in the details of a problem to look at the situation as a whole.

Many people have gotten so sidetracked by the myriad of "bells & whistles" within their various devices, apps and online services that they lose sight of their **primary objective** for using those things to begin with.

This is understandable as baby boomers grew up with specialized, single-use gadgets and machines (which were usually pretty intuitive to use).

Of course, even single-use household appliances have to be used *methodically* (e.g. you would want to put your clothes into a washing machine **before** turning it on) while paying close attention to visual cues such as status lights.

Indeed, the very names of these appliances clued us in as to what they could be use for...

- ✳ blenders
- ✳ dishwashers
- ✳ toasters
- ✳ can openers
- ✳ calculators

To be sure, there would be various **settings** on these devices that we could ponder and consider using but none of which would seriously distract us from the current task at hand.

This is why **focus** is even more important now than it was when watching a photo come directly out of a Polaroid Instant Camera was a jaw-dropping experience 😯

The Nuts & Bolts of Technology Use

"SAM THE GREAT TIRE CHANGER"

Every time I have to change a tire, it seems Magilla Gorilla was the last one to put the lug nuts on.

I once asked my shop teacher if the lug nuts would come off easier if I greased them.

He said: "They'll come off when you're doing 60 down the freeway."

Painting a picture

People who are the least efficient with computers, tablets, smartphones, wearables and other devices are the ones who have no clear objectives for using them or often lose sight of their task at hand.

Contemporary technology use is like painting a picture 🖼️

This artist is not simply slapping paint randomly onto her canvas. She is focused. She is not trying to race ahead to her finished work. Each brushstroke has a purpose. And please note her *visualized outcome* is directly in front of her.

The painter's brushstrokes are the "real world" equivalents to keyboard presses, screen taps and mouse clicks.

"TUFFY MONET"

"TUFFY'S MASTERPIECE"

In the technology world, even "single tasks" such as *visiting a website* will have multiple steps — but the good news is, those "multiple steps" will often be present within the app or online service itself in the form of *visual cues*:

CAREFULLY ANALYZE VISUAL INTERFACES FOR FUNCTIONALITY

For example, in the above *web browser window*, we see forward & backward navigation buttons ("breadcrumbs"), three colorful buttons (it's important to know what they are for), a "New Tab" label and a web address box.

The procedure...

1. Make a cup of tea 🫖
2. If necessary, *search the web for instructions.* (e.g. what you want to do with your device of choice) .
3. Start the app you want to use (which you may have come across by *noodling around* your device).
4. Think about what you want to do 🤔 and then look around for a way to do it 👀
5. Breathe (you need to be calm and relaxed).
6. Remember to proceed in a *logical manner* 🤔
7. Keep your objective firmly in mind 🧠 (this will help you "filter out" things that you currently have no use for).
8. Apply real world commonsense and methodologies (for example, browsing around your device for a particular *file* using the technology basics is **conceptually** just the same as searching through a card file box or a filing cabinet 🗃️ 🗄️).

And it helps a lot to understand **conceptually** what you want to do before you reach for any of your devices.

For example, putting a photo in an envelope and then mailing it, corresponds to attaching a photo to an email message and then sending it:

Despite their digital nature, contemporary technology-related tasks will often have some real world correlation.

And then...

Regardless of which app you are using and which device you are using that app on, simply think about what you want to do and then look around for a way to do it...

The only thing left to do is to send this email (again thinking in terms of logical procedure):

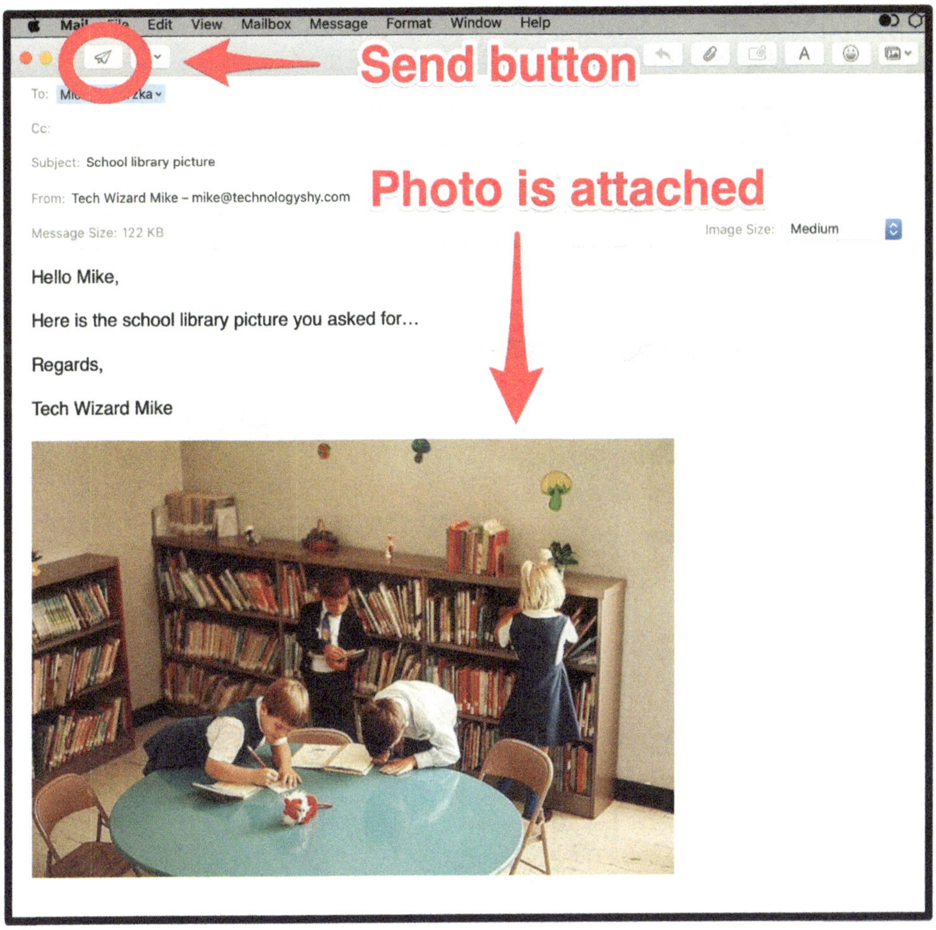

Teaching you how to fish!
Appearances will vary widely between applications and online services. As pictured above, look for things that you can use to accomplish your current task at hand.

GIGO

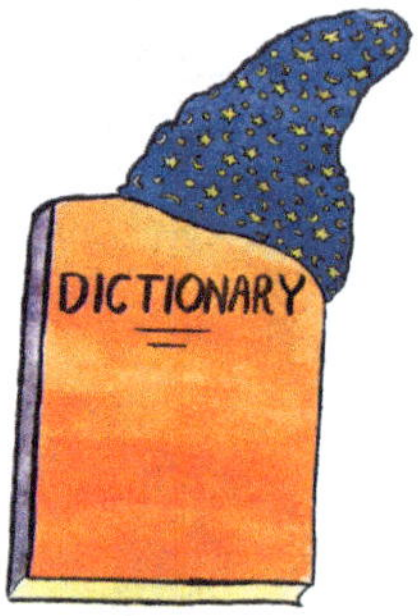

GIGO Stands for "Garbage In, Garbage Out." GIGO is a computer science acronym that implies bad input will result in bad output.

GIGO is as true today as it was in 1981 when your Tech Wizard Tour Guide's first assignment in his "Introduction to Computer Science" class was to get a mainframe computer to print his name through the use of punch cards.

(Pre) Tech Wizard Mike wasted **a lot** of punch cards trying complete that assignment due to his impatience, lack of methodical procedure and not paying particularly close attention to what he was doing❣️

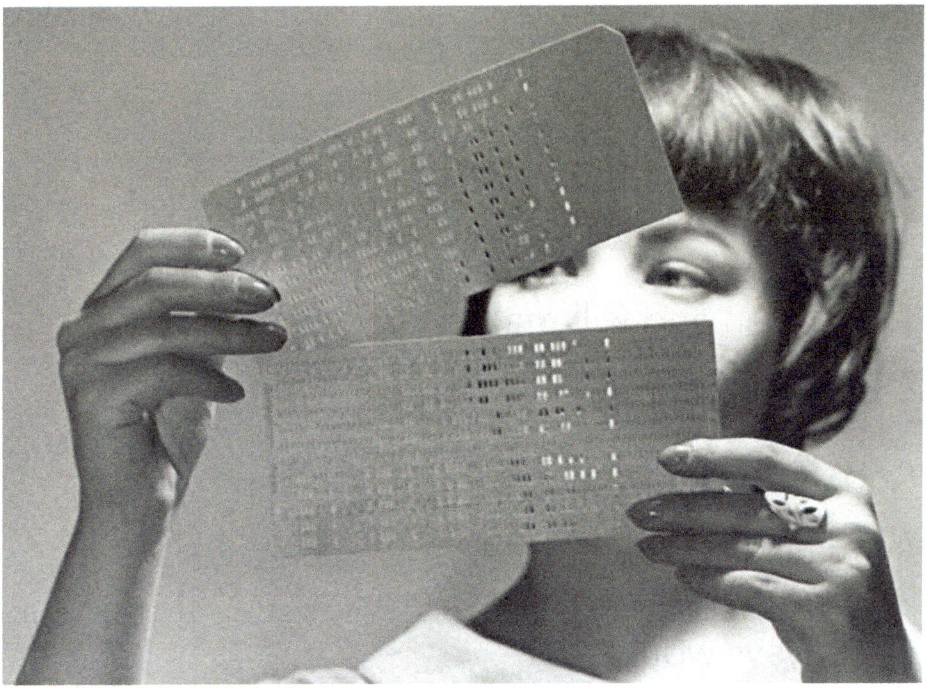

THIS IS WHEN MY TEACHER TAUGHT ME THE GIGO PRINCIPLE 👩‍🏫

Expect the Unexpected 😯 😒

"TRAFFIC LIGHT AND PUZZLED TUFFY"

Things can and will **malfunction** in the technology world just as they do in the real world.

And, in the technology world (just like in the real world), things do not always work as they should — or as we would expect them to.

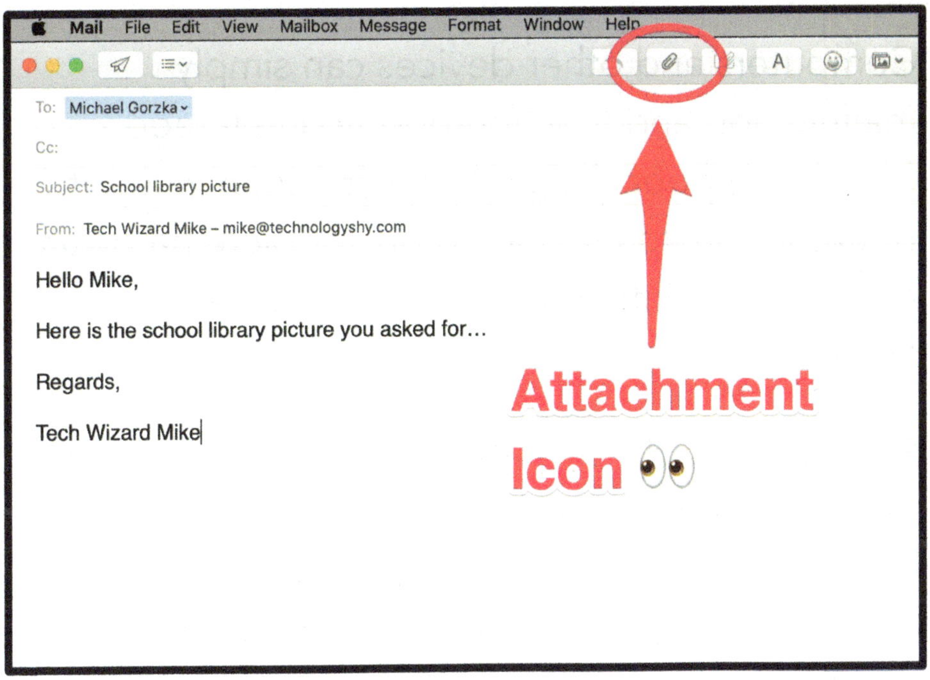

LOOK FOR VISUAL CUES SUCH AS ICONS & BUTTONS 📎

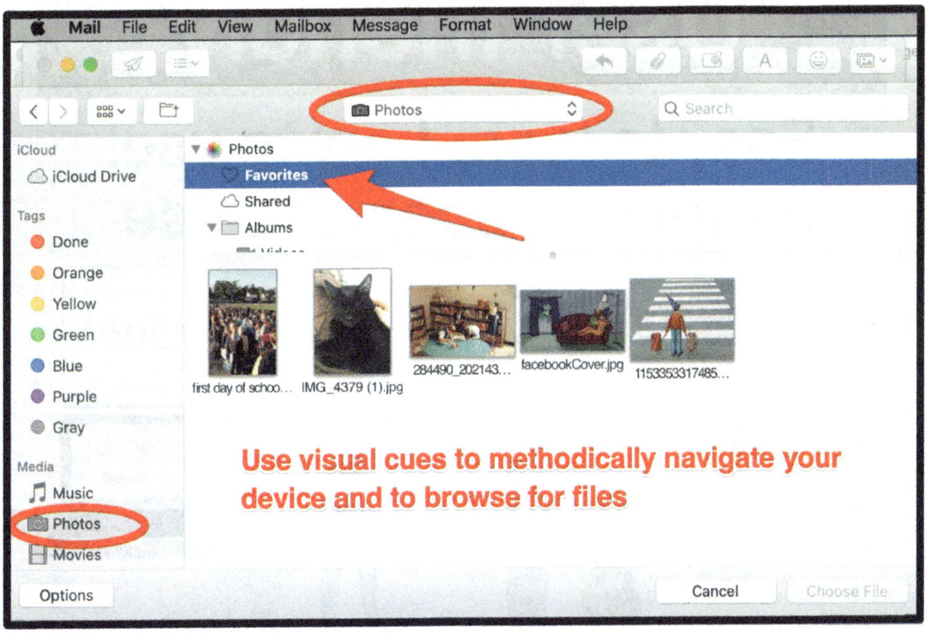

USE VISUAL CUES AND THE TECHNOLOGY BASICS TO METHODICALLY NAVIGATE YOUR DEVICE AND TO BROWSE FOR FILES.

Computers and other devices can simply malfunction and function poorly due to age, *bloatware*, poor design & construction, current *operating systems* running on *obsolete hardware* and malware infection (commonly called "viruses").

People who administrate **websites** (which contain online services & resources) often make poor decisions resulting in websites that are dysfunctional and/or not user-friendly (i.e. not intuitive).

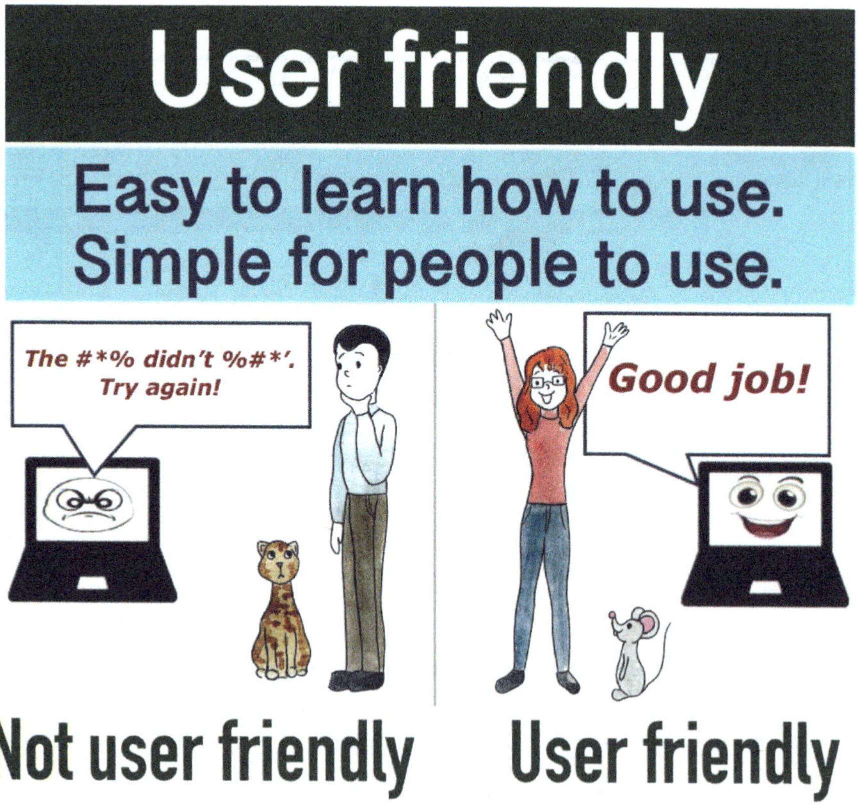

"THE BAD COMPUTER, THE GOOD COMPUTER"

And also, many website administrators are often simply out of touch with the needs of their users.

AND/OR THEY ARE OVERWORKED 😰

THEREFORE IT IS VERY IMPORTANT TO STAY FOCUSED ON YOUR CURRENT TASK AT HAND AND GET SOME QUALITY HELP IF NEEDED.

"Coders" 🤓

People who have been tasked with creating websites and other online tools & resources are flawed vessels (as all humans are) who are often working under less than optimal circumstances — resulting in **poorly designed websites** (i.e. challenging to use and to navigate).

Although to be fair, this woman's laptop looks to be pretty long in the tooth (possibly even *obsolete*) which could be the reason why her workflow has been derailed.

Suggested procedure (workflow) for making productive use of any website or app...

- Make a cup of tea (at least visualize the logical & methodical tea brewing procedure).
- Breathe.
- Remember that *haste makes waste.*
- Proceed methodically.
- Scroll up and down (make sure you see everything).
- Examine the interface; think in terms of *functionality.*
- Use the technology basics to click and tap your way around the interface but do so with thought and intention.
- Use forward & back buttons
- Look for menus.
- Look for buttons.
- Look for sidebars.
- Summon *context menus* 👩‍🍳
- Get good help if you need it (and *take notes*).

When anything unexpected (or simply new & unfamiliar) presents itself…

- Come to a full stop ✋
- Do not click or tap or type a gosh darn thing.
- Sip your tea ☕
- Simply take it in what you see 👀
- *Breathe* (it's important to stay calm).
- Calmly analyze whatever it is that gives you pause.
- Look for notifications and/or error messages and document them.
- Search the web for those error messages 🔍
- Do not fall for *scams* (e.g. fake virus alerts).
- Do not provide **any** personal information.
- Get *good help*.

Keep in mind that every day technology use will often be like Wednesdays on the old Mickey Mouse Show...

"A TYPICAL WEDNESDAY"

During the middle of the week, the Mickey Mouse Club decided that Wednesday was "Anything Can Happen Day" or as I like to call it: "No One Knows What the Hell's Going On Day." — Wade Bradford

Practice Squeaky-Clean Technology Use

- Use commonsense (e.g. If something seems to good to be true, it will be.)
- Keep your web browsing squeaky clean (e.g. stay away from websites that offer such things as gambling, online games, coupons, free movie downloads and celebrity gossip).

- Rule of thumb is to not download apps from outside your device's App Store (*unless you really need to do so*)
- Do not try to download any free music or movies that are copyrighted.
- Only open email and message attachments from people you know and trust (and even **then** we need to be wary about attachments due to *spoofing*).

Better Safe Than Sorry 😐

Back up important files using secure *cloud services* **and** *physical storage devices* (while knowing that such cloud services can be discontinued and storage devices might arbitrarily turn into pumpkins).

Archive important email and text messages (apply the *do it, defer it, delegate it, ditch it, archive it*[6] method).

You can also *email important materials* to yourself and then archive them.

You can *digitize paper notes* (via a smartphone scanning app).

Visit www.technologyshy.com/videos for tutorials on the above back up methods — and more, including how to download apps (if necessary) from outside your device's App Store.

[6] Mike, Tech Wizard. "Do It, Defer It, Delegate It, Ditch It, Archive It." *Help for the Technology Shy*, www.technologyshy.com/inboxzero/.

Search for It

"SEARCHING HARD"

Whether you are preparing for a task or a project... 🔍

Or for general interest... 🔍

Or for future reference... 🔍

Or if there is something that you unexpectedly need to find out as you are diligently working on your present task at hand... 🔍

You can search the Web!

In a nutshell... 🐿️

Know which device you are using; know which app or online service you are using; know what you want to do or find out → and then search the web 🔎

How do I create a presentation on an iPad?

Is an iPhone 4 SE obsolete?

Can you use Skype on a Windows 7 computer?

How do I back up files on my MacBook Pro?

What are Windows 10 keyboard shortcuts?

On any internet-connected device...

1. *Noodle around* for your device's web browser.
2. Type what you are looking for into the web address / web search box.
3. Click or tap the "Go" button (or press the enter / return key on your computer keyboard).
4. Peruse the search results page.
5. Click or tap on promising web page links.
6. Use "breadcrumbs" (e.g. back and forward buttons) for navigation.

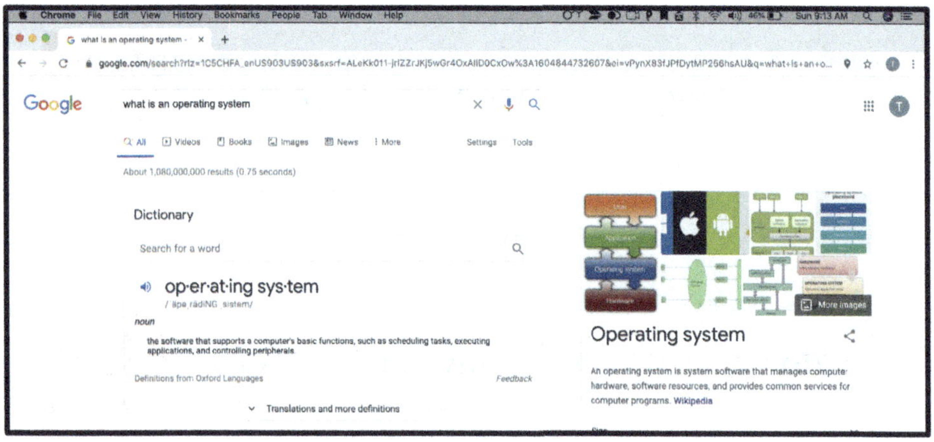

SEARCH THE WEB FOR UNFAMILIAR TECH TERMS (AND ANYTHING ELSE YOU WOULD LIKE TO KNOW).

Search your device

It is important to have what you need when you need it.

You can search and/or browse for files, apps and *data* (e.g. contacts and calendar events) that are stored on your device.

You can search and/or *browse* for things that are stored in your *cloud service* (assuming your device has an app for your cloud service and you are signed into that cloud service).

Where is your device's search box?

Once again, look for *visual cues*; on a desktop or notebook computer, simply look for something like this…

On a mobile device, you may need to use *finger gestures* to reveal the search box.

Exactly **which** finger gestures will depend on the *operating system* that is running on your device (yet another reason for *knowing what you are using*).

Simply use your device (or another internet-connected device) and perform a web search:

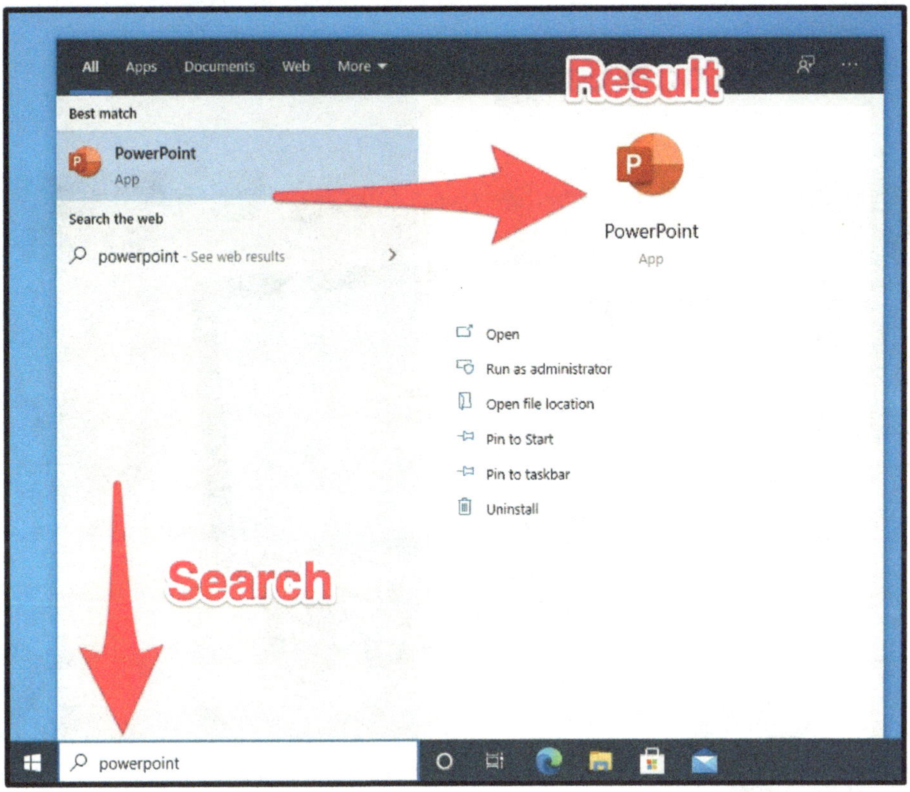

SEARCHING THIS COMPUTER FOR THE "POWERPOINT" APPLICATION

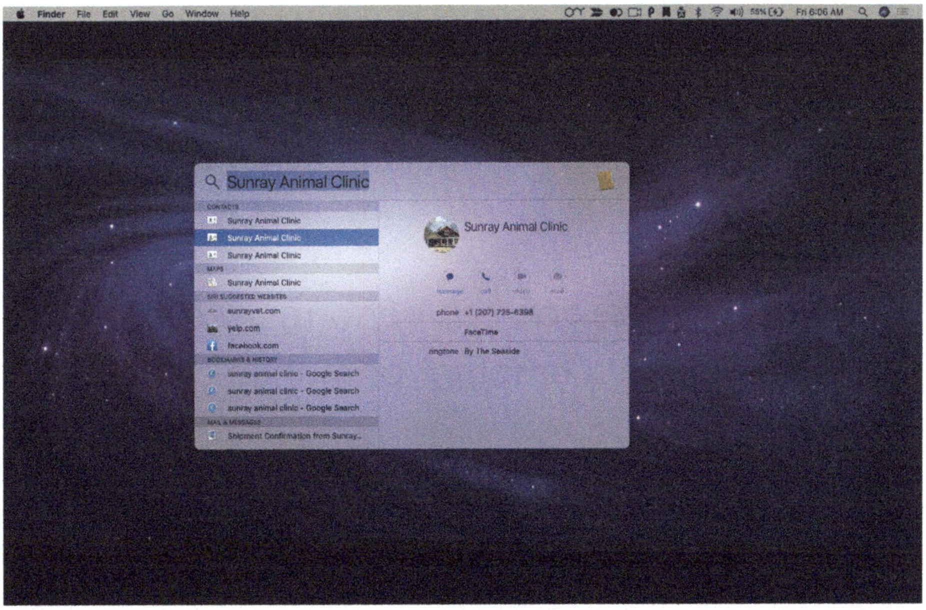

SEARCHING THIS COMPUTER FOR CONTACT INFO

Good Grief, What Is This!?

* Bloatware? (In this example, **no**.)

* A scam? (In this example, an emphatic **yes**.)

* A legitimate system notification? (In this example, **no**.)

If you're not sure, you can perform a web search (using a different computer or mobile device).

And if you are still not sure, you should get some *good help*.

Beeee careful reminders 🐝

- Any pop-up message on a web page (*context matters*) informing you your device is at risk (or words to that effect) will be a **scam** 😈
- Any pop-up message on your screen that is threatening will be a **scam** 😈
- Any pop-up message demanding that you pay money or divulge your confidential information will be a **scam** 😈
- Any pop-up message providing a phone number will **scam** 😈

THE SHORT ANSWER IS: DO NOT ENGAGE!

Uh, Oh... 🛸

“TIN FOIL CAPS BAND”

Nothing is more disconcerting (except perhaps an "Earth vs. the Flying Saucers" scenario) than a cryptic and/or ominous message that suddenly appears on your screen.

Keep calm, make a cup of tea and evaluate ☕

Is it legitimate? Is it a scam? Is it a sales pitch? Is it **really** a message from the FBI saying "Pay up or else!" (Spoiler alert: It won't be 🙂)

Sources of threatening (or at least disconcerting or annoying) messages include:

- system messages (no problem 😌)
- bloatware (no problem but can be really annoying 😒)
- website scams (no problem but do not engage 🤔)
- web browser shenanigans (can be a problem, so get good help 🧙)
- device infiltration (problem‼️🙀)

Threatening messages (in general)

"Ransomware" could be a website bluff or an email threat or it could seize your entire device.

In any case, **do not engage** 🛑

Do not pay them anything.

Do not communicate with them.

Do not provide any confidential information — or indeed any information at all.

Did I Mention That Context Matters?

"The miserable"

Legitimate system messages

It is important to identify the *source* of the message.

Is it coming from a web page you landed on? Or is it a *system message*?

Legitimate system notification messages will simply inform you.

They will not try to scare you or coerce you into buying anything.

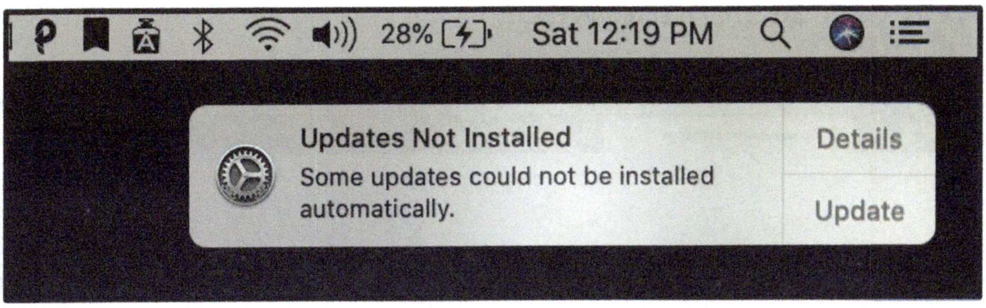

LEGITIMATE MACOS UPDATE NOTIFICATION

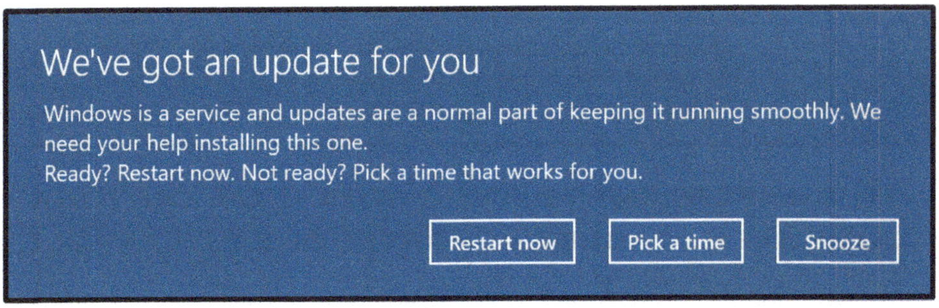

LEGITIMATE WINDOWS 10 UPDATE NOTIFICATION

Legitimate system messages (i.e. messages from your device's *operating system*) often convey importance and will announce the availability of **security updates** for your device.

But they will never be threatening or intimidating.

It is very important to know what your device's system messages are supposed to look like.

If you are not sure what they are supposed to look like, armed with the knowledge of which device you are using and/or which OS is on that device, you can *search the web*.

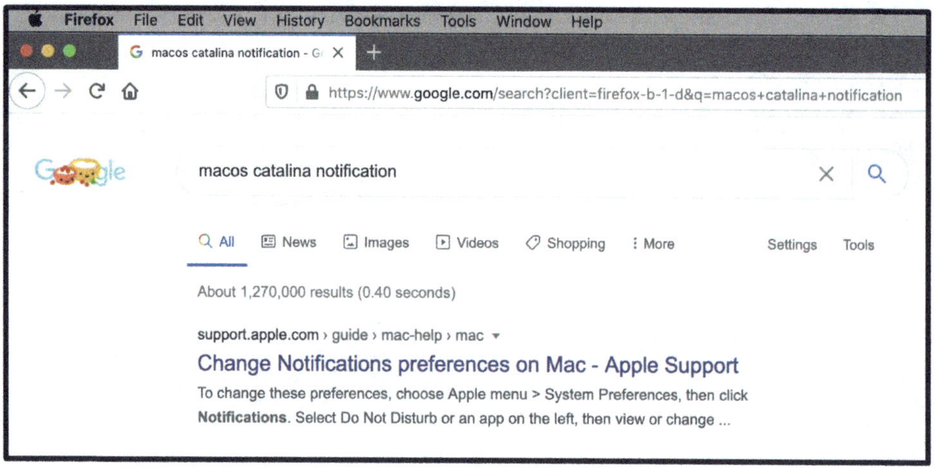

IT IS IMPORTANT TO KNOW WHICH OS ON YOUR COMPUTER / TABLET / SMARTPHONE / WEARABLE

Virus Detected!

ILLUSTRATION BY DINORA NURTDINOVA

1. First of all, any "virus detected" messages presented as in the above illustration are going to be a scam. **Do not engage**. If necessary, simply power off your computer or mobile device.
2. Secondly, practicing the *Beeee Careful Guidelines* are going to be far more effective (and cheaper!) than any "virus protection" software.

Bloatware

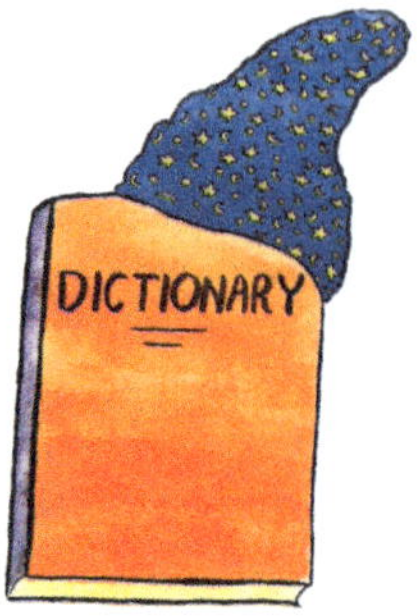

Bloatware is a slang term for unnecessary programs that are pre-installed on some new computers (often making them slow and unwieldy).

"Bloatware" comes preinstalled on some devices (usually in the form of a trial version) in a business partnership between the device manufacturer and the antivirus software company.

"KRAMDEN"

It is important to show bloatware the door so to speak...

"NOT WELCOMED"

...because bloatware will "gobble up" the resources your computer or mobile device needs to operate properly.

"GOBBLE-IT-ALL"

Once again it comes down to knowing what you are using and having a clear objective.

1. Write down the name of the bloatware (e.g. "Kramden 1955" in our fictitious example).

2. Find out which operating system is running on your computer or mobile device.

3. Search the web for how to remove the bloatware app.

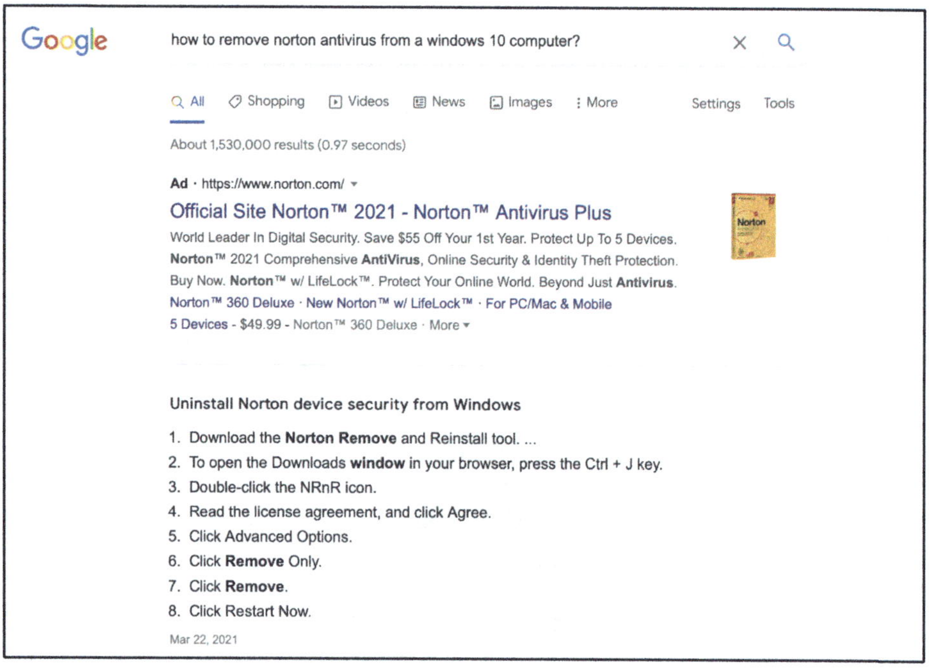

THESE THINGS CAN BE TRICKY! REMEMBER TO DO YOUR OWN RESEARCH; DRAW YOUR OWN INFORMED CONCLUSIONS AND SEEK OUT GOOD HELP IF YOU NEED IT!

Web Page Scams

It is a common enough scenario, you are surfing the web and you are confronted with a "your device is at risk" warning or an even more disconcerting *ransomware* demand 🦹

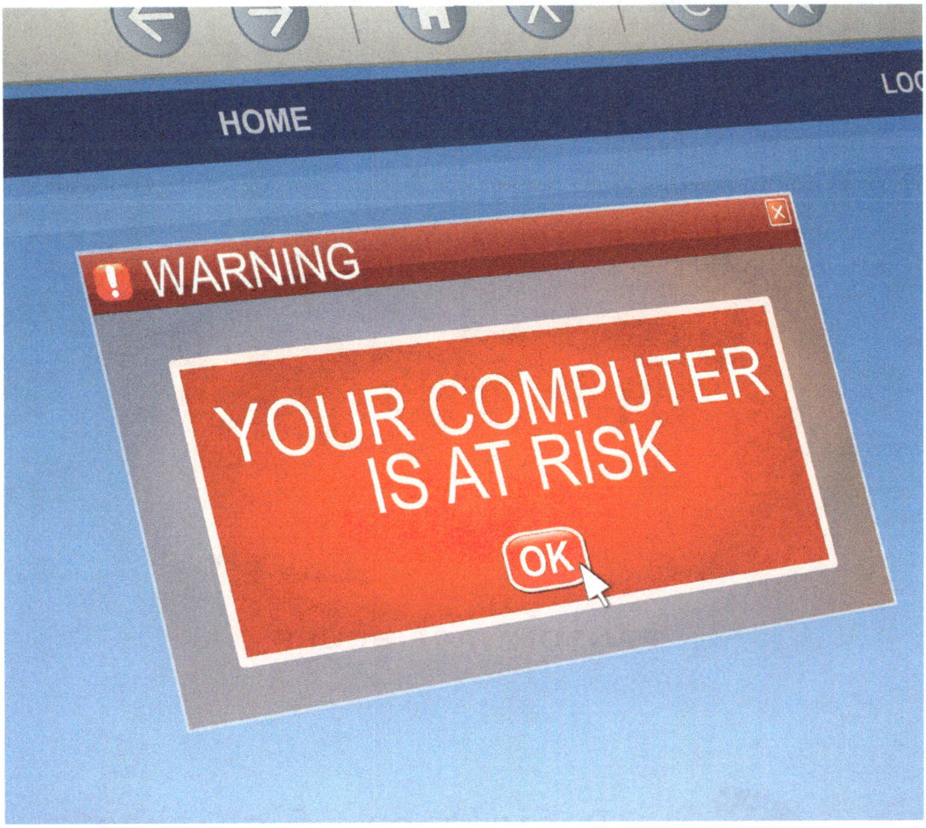

THESE THINGS CAN BE TRICKY! DO NOT ENGAGE!

Do not click or tap anything on such messages.

Do not provide any personal information or indeed any information at all.

(Even if the web page was able to take your picture using your device's built-in camera 📸)

RANSOMWARE BLUFF

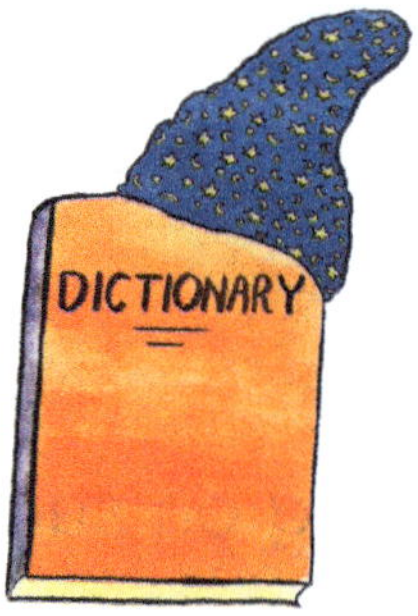

Definition of chicanery

deception by artful subterfuge or sophistry : TRICKERY

What to do if you feel threatened or coerced:

1. Write down the specifics of the threat (so that you can *search the web* for info and/or get *good help*).
2. Quit (or *force quit*) your web browser app.
3. *Clear your web browser's cache* via a *menu selection* or through *settings* (depending on your device).
4. If necessary, power off your device.
5. If the threat persists, get *good help*.

If you are experiencing a lot of pop-up ads, it may due to the presence of *adware* on your device.

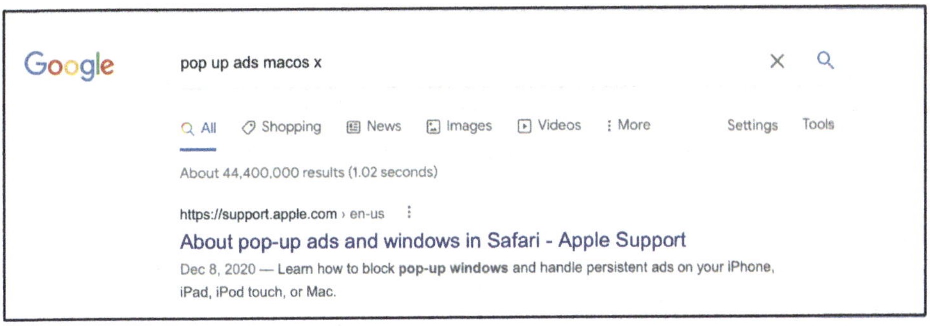

Obsolete devices are especially vulnerable to *malware*.

Does Your Device Already Have Virus Protection?

Your device may be "naturally" (so to speak) less vulnerable than other devices from viruses and other forms of malware; **or** virus protection software may already be present on your device.

For example: "Windows 10 provides comprehensive, built-in protection—at no extra cost."[7]

Search the web for your operating system's *built-in virus protection* 🔍

Find out what those messages **are supposed to look like** on your device.

If your device's *OS* has built-in virus protection, you will not have to pay for it.

Any message that asks you to pay for virus protection will be a scam or *bloatware*.

---

[7] *Windows Security: Microsoft Defender*. (n.d.). Microsoft. https://www.microsoft.com/en-us/windows/comprehensive-security

People often ask your favorite Tech Wizard Tour Guide...

"Which virus protection products do you personally use?"

To which my answer is... "**None**" 🤯

Your author is crazy and believes in UFOs 😜

Actually he feels pretty sane at the moment, thank you 😉

(And he keeps an open mind in regards to UFOs 🛸)

A Bold Experiment

At the time of this writing (Summer 2021), the latest version macOS **does not** have "built-in virus protection" per se and the latest version of the Windows OS **does** (i.e. "Windows Defender").

Your author dragged his macOS notebook computer and his Windows notebook computer through the cyber-mud so to speak → visiting an assortment of gambling, celebrity gossip and yes, adult entertainment websites 🙈

And the result? No viruses or any other form of malware-infection on either notebook computer.

However, I did not fall for any *scams* nor did I install new apps on either computer from outside each computer's *App Store* nor did I do anything **illegal** nor did I use *weak passwords* and was very careful with message attachments 📎

Possible signs of device infiltration

- Your friends are getting strange emails from you.
- Your device runs very slowly.
- Frequent pop-up windows appear, especially the ones that encourage you to visit websites, or download *antivirus software*.
- Frequent crashes or unusually slow computer performance.
- Your device does things on its own.

How to avoid device infiltration

There is frankly no substitute for *common sense* (to avoid falling for *scams*) and *squeaky clean* technology use.

"SOAP BATHING"

Factory Reset

If you feel your device has been compromised, your Tech Wizard Tour Guide's best counsel is to...

1. Back up your files & data.
2. Use another device to change all of your website passwords.
3. Record the new passwords into your technology notebook (which already **has** your device's admin usernames & passwords).
4. *Factory reset* your compromised device.

"THE GREAT WELDER"

A Factory Reset removes all user data from a device and returns it to the way it was when you first purchased the device.

Data like contacts, photos, apps, your cache and anything else that you saved onto the device ever since you started using it will be cleared from it.

"FACTORY RESET NOTEBOOK WITH SOAP"

Virus alert messages can be the result of annoying, device hobbling, productivity decreasing *bloatware* → and factory resetting your device can sort that out as well.

Five quick tips for moving forward

After factory resetting a device...

1. *Beeee careful* 🐝

2. Keep your device's *system software* up-to-date (as that software will include security updates).

3. Practice *technology minimalism.*

4. Keep backing up your files.

5. Always proceed mindfully toward your clearly defined goals.

Cool as a Cucumber

This is what you need to be during technology use:

Cool as a Cucumber

"AS COOL AS CAN BE"

Regardless of any apps or devices or built-in virus protection, you need to be *calm, cool and collected* during technology use. To be otherwise would be inviting lost work & time, additional frustration and malware infestation.

Good Help Is Not Hard to Find

Once you have your *technology foundation* in place, getting help with whatever you want to do can be pretty gosh darn easy.

You just have to know…

- which device you are using (e.g. a computer running the Windows 10 operating system)
- which application you are using on that device (example: Microsoft Word)
- what it is exactly you want to do (clarity & focus are keys to productive, stress-free technology use)

Ask a good egg

There are knowledgeable, unpretentious people who will want to help you simply because they are kind and interested in effecting positive outcomes.

And they are your go-to people!

In a nutshell, knowing which device you have and which OS is running on that device, makes it much easier to get good help:

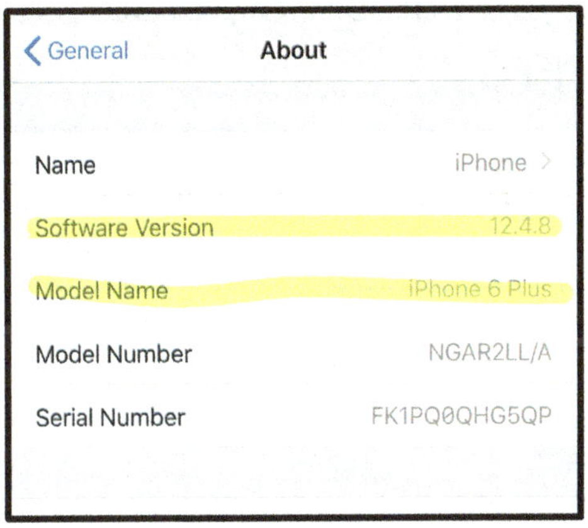

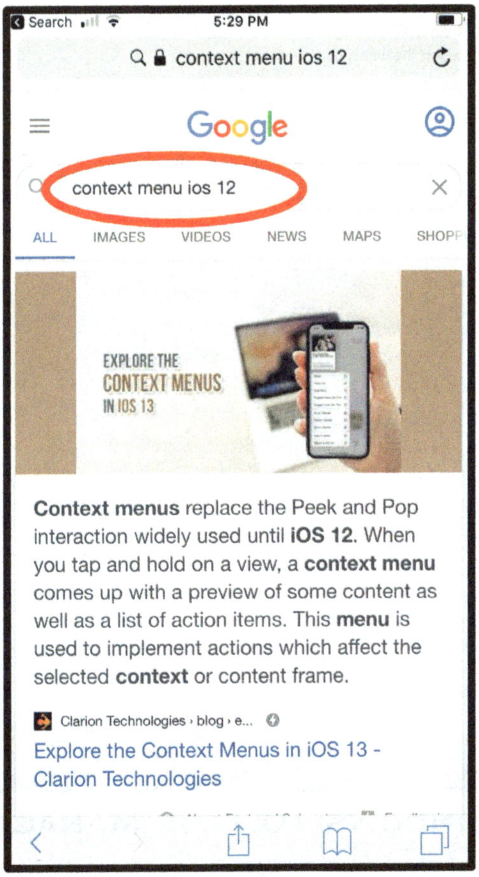

Web search

As we previously discussed, armed with the knowledge of what you are using and the *correct terminology*, you can simply type your question into a search box:

How do I force quit an app on Windows 10?

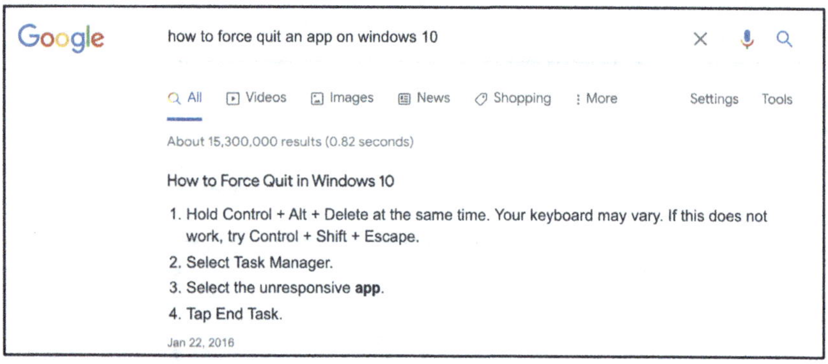

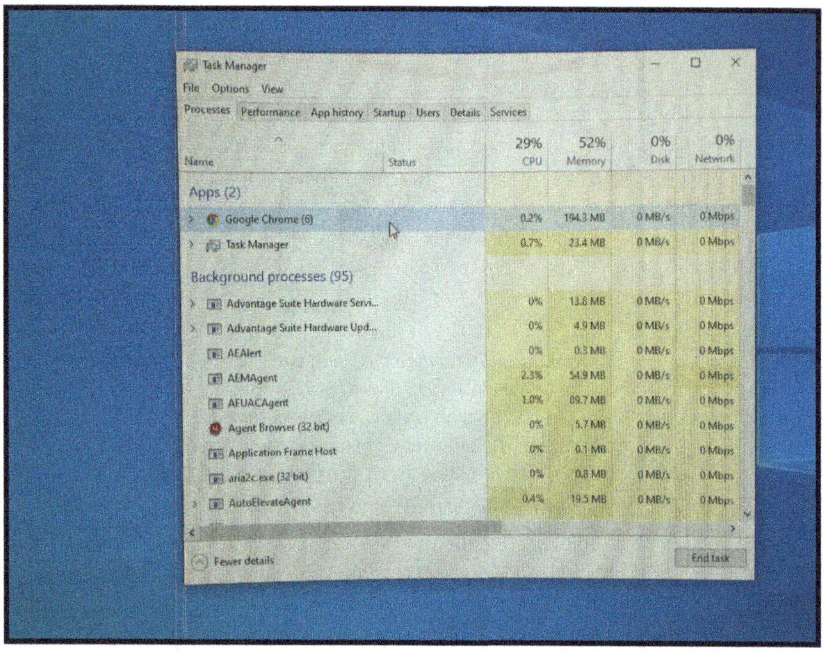

WINDOWS 10 TASK MANAGER

Support forums

If you know what you are using...

EXAMPLE: WINDOWS 8.1 ON HP PAVILION ULTRABOOK

… you can post your question directly to the appropriate **official** support or community forums for your particular device and/or operating system:

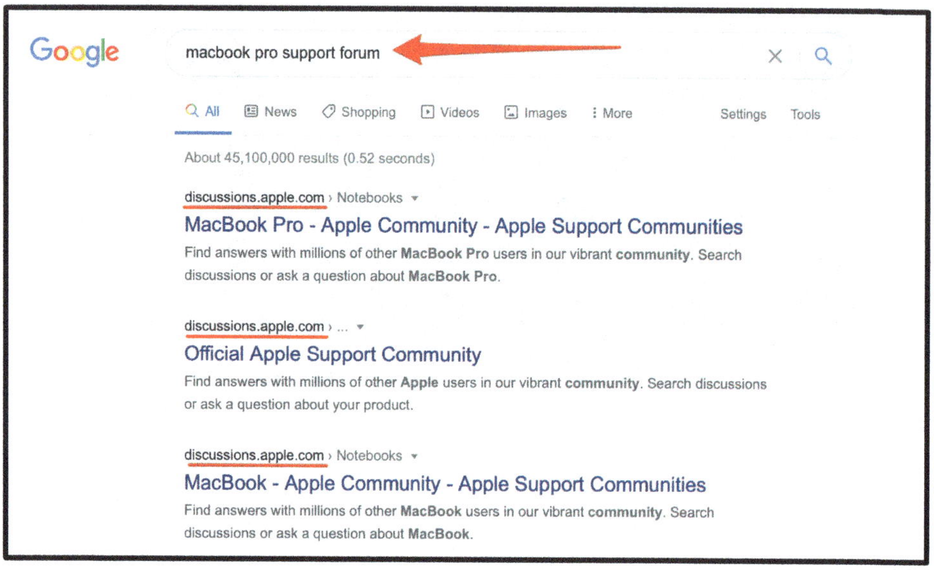

SEARCH BY DEVICE

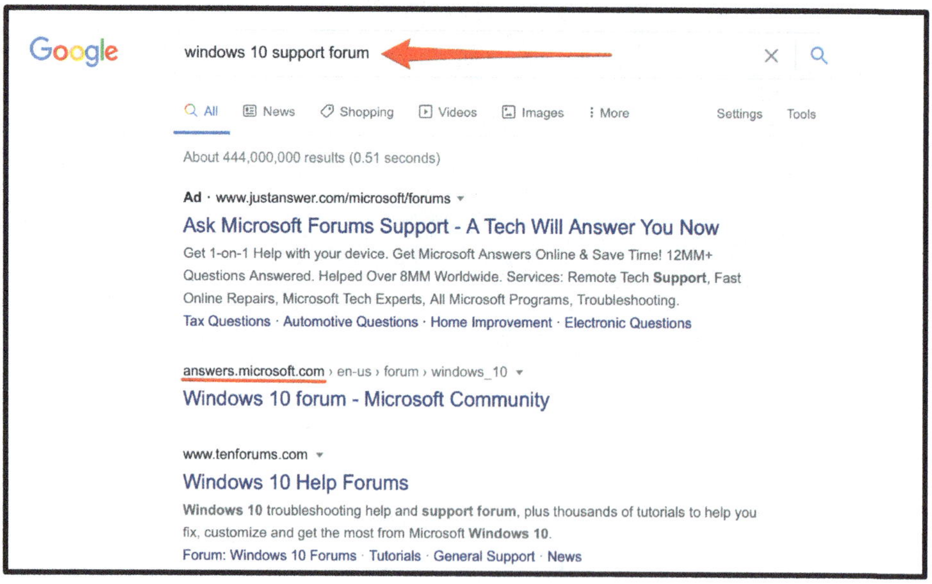

SEARCH BY OPERATING SYSTEM

A device is often **defined** by its *operating system* (e.g. "Windows 10 notebook").

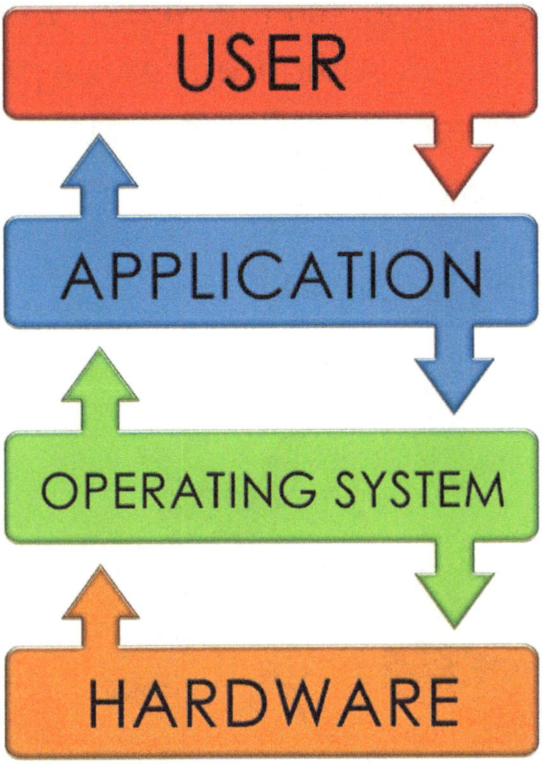

Customer support ☎️

You can perhaps cut to the chase as it were and contact customer support for your device → either through email, chat or phone.

You can search the web or visit your local library for your particular device's customer support **contact info** → and diligently record it into your *technology notebook* 😸

Authorized service providers

If you think your device may not be functioning properly, you can search the web or visit your local library and find an *authorized service provider* → and diligently record their contact info into your *technology notebook* 😸

Well-meaning jackasses

“ANNOYED DEVICES”

There are people who usually mean well but will often merrily pass along misinformation. Well-meaning jackasses can often be identified by their giddiness and their reflexive, off-the-cuff answers (and lack of thoughtfulness).

Know-It-Alls 🧐

These are people who use their knowledge of technology to project their self-importance.

Their primary purpose is to boost their egos — and not to effect positive outcomes.

They can be identified by their assumed superiority and lack of true empathy & thoughtfulness.

“KNOW-IT-ALL”

No-goodniks

There are scam artists out there. There are people who **prey** on trusting souls — and a pox on them!

THEY WILL LIE LIKE RUGS TO GET YOUR PERSONAL INFORMATION

DON'T GIVE IT TO THEM!

Rules of thumb

Remember that no reputable company will ask for your personal information.

Do not do anything that commonsense tells you would be counterproductive (such as sticking a metal paperclip into a smartphone's delicate electronics-filled charging port)

Cross-reference what you are told (especially when the advice is unsolicited).

Take everything a *well-meaning jackass* or a *know-it-all* says with several grains of salt (at least)

Take full-responsibility for your technology use.

Do not surrender *the administration* of your computer / tablet / smartphone to anybody.

Be well-aware of your surroundings if you are in a public space with your *technology notebook*.

The "Sisyphus Syndrome"

"BOULDER PROBLEM"

In your Tech Wizard friend's experience, some people never allow themselves to be **done** with a particular task or project.

For example, I helped a gentleman we will call "Ted" complete a task that was more complicated than it should have been primarily due to the fact that Ted insisted on using his **preferred** apps (which he apparently had some emotional investment in) instead of better options.

Obstinance can be extremely counterproductive.[8]

Ted had indeed completed his stated objective (after getting some *good help* for yours truly) but instead of moving forward, Ted instantly became fixated on why his preferred apps did not work as he thought they should.

Oh my gosh, please don't do this!

Yes, obstacles will appear on your path to task / project completion and you have to be mentally and emotionally prepared to detour around them.

Ted's preferred apps were not the best choice for what he needed to do because they proved to be ineffective.

---

[8] Gorzka, Help for the Technology Shy, 340.

Do not let vague or imaginary issues drag you back to square one.

And you may also need some **willpower** to disengage.

A clear, written or sketched *objective* can help you stay on point. Did you accomplish what you needed to do?

We must always be flexible; accept the flaws in technology, in other people and in ourselves — so that we can concentrate on the more important things in life and move **forward**.

CONTEMPORARY TECHNOLOGY USE CAN BE LIKE AN ABSTRACT JIGSAW PUZZLE; SEEK OUT POSITIVE COLLABORATIONS

Practice Technology Minimalism

CHARLES M. SCHULZ BELIEVED THAT THE KEY TO CARTOONING WAS TO TAKE OUT THE EXTRANEOUS DETAILS AND LEAVE IN ONLY WHAT'S NECESSARY.

We should apply this same philosophy to our contemporary technology use.

KISS = "Keep it simple, snookums"

In 1984, this cute little guy was the *ne plus ultra* of *usability*:

THE ORIGINAL APPLE MACINTOSH COMPUTER

To be fair there were only a handful of things you could do on this computer — as it shipped with a word processing app, a drawing app and some games.

However, if you keep your contemporary technology use focused and specialized, then your technology use can be as easy peasy lemon squeezy as things were in the mid to late 1980s.

Today, there are a *plethora* of devices and peripherals available to us.

“DEVICES”

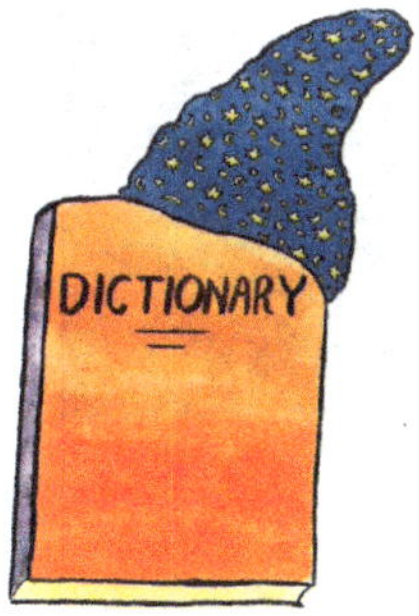

A plethora of something is a large amount of it, especially an amount of it that is greater than you need, want, or can cope with.

Many of my technology shy friends have purchased devices they did not need based on the unscrupulous (or perhaps just tech-happy) advice of sales people, *well-meaning jackasses* and *know-it-alls*.

"Jackie" did not feel comfortable and proficient with the tablet's onscreen keyboard (she needed the *correct terminology*) which necessitated the purchase of an *external keyboard*.

peripherals query

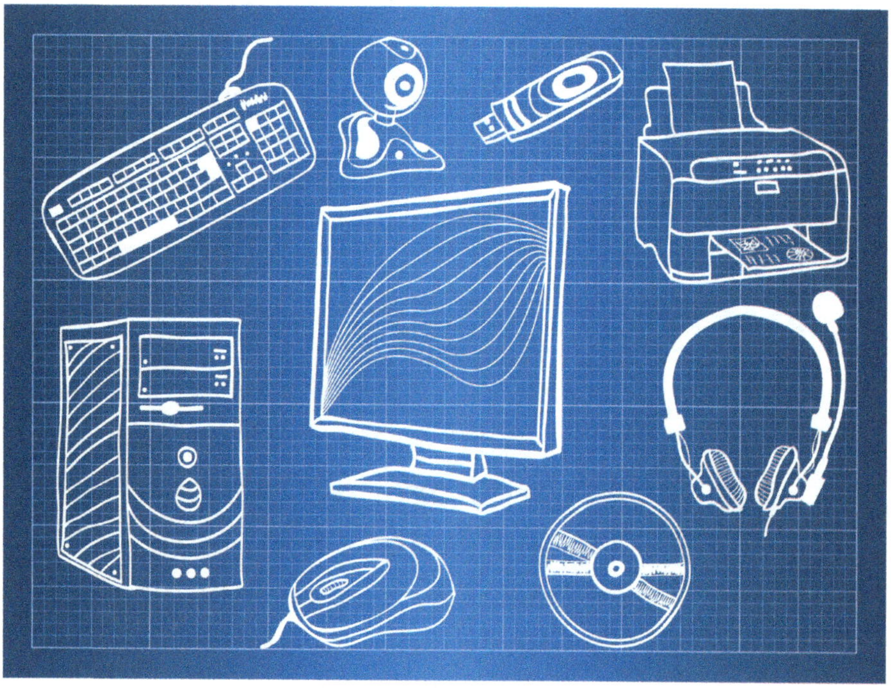

Peripherals (or "accessories") are things that can be physically attached or wirelessly connected to a computer / tablet / smartphone for a specific purpose (e.g. a flash drive to backup files).

Compatibility will often be an issue between **peripherals** (e.g. mice, trackpads, keyboards, printers and speakers) and **devices** (e.g. computers, tablets and smartphones).

Keyboards will have often have **special function keys** that will be specific to a particular device (or more accurately the operating system or "OS" that is running on that device).

SPECIAL FUNCTION KEYS POINTED TO ABOVE

How many of these do I actually need?
What would you like to do?
Email and looking things up online.
Which of these devices do you feel most comfortable with?
This one. But I do not like using the onscreen keyboard.

How do you know which peripherals are compatible with a particular device?

Simply search the Web 🔍

This is one of the many benefits of knowing what you are using.

1. Identify the potential peripheral (i.e. brand and model).
2. Gather your device's info from your *technology notebook* (i.e. brand, model and OS).
3. Formulate your compatibility query and search.

Can I use an Apple Magic Keyboard with a Samsung Tablet?

"There are three things I have learned never to discuss with people... Religion, Politics, and The Great Pumpkin." — Linus Van Pelt

"SUPER PUMPKIN"

It should be noted here that *third party peripherals* (i.e. peripherals that were manufactured by a different company that manufactured the computer / tablet / smartphone / et al.) seem to have a greater potential for arbitrarily turning into pumpkins → so please make sure there is a warranty before you buy one.

Pondering accessories 🤔

There are going to be **a lot** of options out there — some of which would actually decrease your productivity by making your device harder to use.

For example, many of my clients have purchased cheaply made (but expensive 🤑) plastic cases that were unwieldily and even made their beautifully designed tablets and smartphones look "fugly" 😒

Here are four guidelines...

1. Stay away from anything plastic.

2. Look for ease of use.

3. Be prepared to pay more for quality.

4. Think in terms of increased productivity.

Nothing is engraved in stone here 🗿 Evolving hardware, peripherals, apps and services can provide a variety of options → the key is to stay focused on your objectives (and not lose sight of the *forest through the trees*).

Let's Beeeee Careful Out There!

"SCAMS EVERYWHERE"

"DONKEY"

And be very careful with attachments, my friend…

“ENVELOPE AND SMILEY”

Guidelines

- Stop, look and think before you click or tap.
- Practice *squeaky clean web browsing* (e.g. stay away from websites that offer such things as games, gambling and free movies & music downloads).
- Know that *email attachments* can be especially fraught with peril 📎[9]
- Keep in mind, if something seems to good to be true, it most likely will be.
- Cross-reference things you are not sure about (e.g. contact UPS by phone if you get an email about a package that you were not expecting) and what you have been told by *well-meaning jackasses & know-it-alls.*
- Practice *technology minimalism*; "gadget lust" is counterproductive.
- Download any new useful apps from your device's *App Store.*

---

[9] Gorzka, Help for the Technology Shy, 319-338.

- Use your **real-world common sense** 🤔 (for example, in actuality, UPS would either deliver a package 📦 or leave you one of those slips (i.e. "first delivery attempted," "second delivery attempted," "third delivery attempted" and finally, "come and pick it up or it's going back!").
- Do not buy additional virus protection products (only not falling for scams and squeaky clean tech use can protect you).[10]
- Do not download any products that promise to "clean up" or "protect" whichever device you are using → it is better to do this "*manually*" so to speak.
- Take **responsibility** for your own technology use; do not outsource any of it.
- Whenever you are threatened by anybody (including **purported** Law Enforcement), get good help.[11]

---

[10] Gorzka, Help for the Technology Shy, 210-215.

[11] Gorzka, Help for the Technology Shy, 319-338.

Technology Use on the Job

Preparation (conceptually speaking) for productive, stress-free technology is closer to paper airplane folding than rocket science.

"ROCKET SCIENCE"

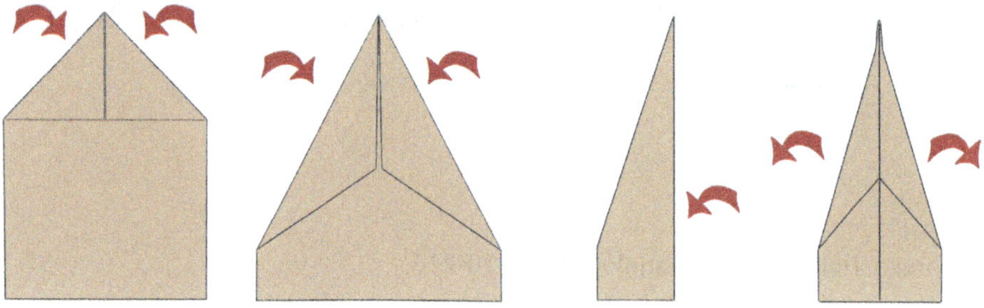

"PAPER PLANE MANUAL"

Three key components

Generally speaking, there are just three (3) key components to productive, stress-free technology use in the workplace.

1. Documentation
2. Breadcrumbs (& other visual cues)
3. Getting good help

PREPARATION FOR PRODUCTIVE, STRESS-FREE TECHNOLOGY USE LOOKS SOMETHING LIKE THIS!

Technology preparation for job

- Get a *technology notebook.*
- Sketch out your job duties & responsibilities → write down anything you need clarification on.
- *Noodle around* with the apps that you need to use → write down anything you are uncertain about.
- Familiarize yourself with any devices & peripherals you need to use. (Search for the device's *About info* and write down the OS or "software version" that is running on the device.)

Any task can seem overwhelming at first...

But take the technology components out of the picture for the moment and take a detached, overhead, bird's eye view of each of your job duties 🦉

This is where you need to mindfully start 🐢 🏁

Visualize your goals and what you need to accomplish 🙌

Work on **understanding** how the various components of what you need to do fit together.

“SAM THINKING”

It is very important to understand **conceptually** what you need to do **before** you reach for any of your devices.

And **then** get your various job components down on paper...

"SAM AT THE DESK"

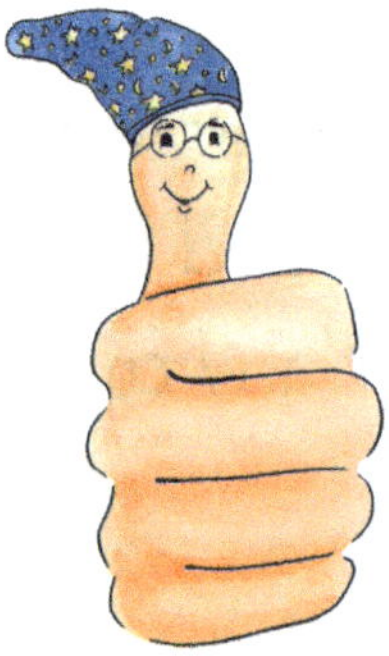

Break down any large task into smaller, more manageable pieces.

Carefully and consciously look at each "piece" or component 🤔

If you are unsure about anything, ask for **clarification** (and record it into your *technology notebook*).

THIS IS WHAT IT'S ALL ABOUT!

CALMLY WORKING TOWARD A VISUALIZED OUTCOME ON A NON-OBSOLETE DEVICE WITH A CUP OF TEA AND YOUR TECHNOLOGY NOTEBOOK CLOSE AT HAND.

In a nutshell, when you are given a technology-related task by your employer...

- Know that mental clarity is crucial. Make sure you understand what you need to do **before** you reach for any of your devices.
- Write down or sketch out what you need to do.
- Break the task at hand into **smaller pieces** if it seems overwhelming.
- Write down any pertinent info that you will need to accomplish this task (or each piece of the task).
- You can **then** use a designated app and/or an online service to carry out that task by methodically following *visual cues* such as icons, buttons and menus.
- Get *good help* if you need it.

Do not psych yourself out 😟 You can do this 👍

Things may **seem** overwhelming at first (or at second third or fourth) but once again my friend, this is by no means rocket science 🚀

Verbal input 🗣️

Toolbox icons, menu selections and navigation breadcrumbs are très fab and can often guide you to a positive outcome with your current task at hand.

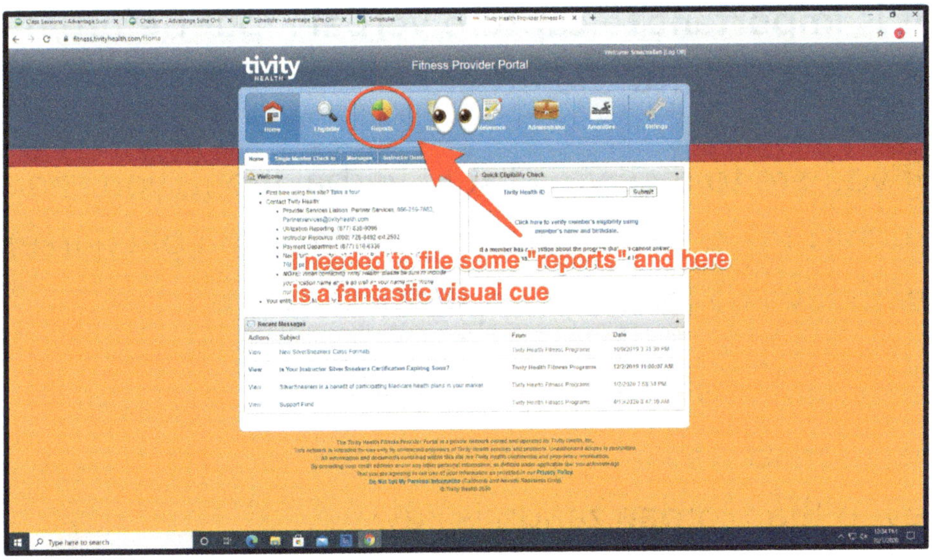

The key is to be *focused, methodical* and *cool as a cucumber* while you are using any application or online service.

For our example, I asked for some verbal guidance 💬 (which I diligently took notes during) and I also used my smartphone 📲 to take some pictures of the app screens showing the various stages for that particular assignment → and reviewed them on a daily basis.

I also took the time to understand **conceptually** 🤔 what needed to be done 🙌 (as simple *rote memorization* and *step-by-step instructions* will often lead to less than favorable outcomes).

It is important to capture verbal input before it disappears into the ether.

If you are to receive **a lot** of verbal input, you can record the person (with their permission of course) with a voice recorder app 🎙️

You can also do things the old fashioned way and take notes but even then, organization is key.

You may find a *designated technology notebook* for your job to be very beneficial.

Proceed methodically

However which way you choose to process the input you are given, you can then take whatever information you have gleaned and then use a designated app to translate that data into the desired result by mindful using *tools* within the app or online service.

Examples of these tools (whose appearances will vary widely between apps, devices and online services) are:

- Buttons
- Icons
- Links
- Menus

The key is to know exactly which result you wish to bring about and to keep that objective firmly in mind as you are using the app or online service / website.

Our "*Think about what you want to do and then look around for a way to do it*" mantra will always be in effect.

Summary

1. Brew a nice cup of tea ☕
2. Document / sketch your job requirements (get clarification if necessary).
3. Peruse the device(s) you need to use (familiarity → confidence → proficiency).
4. Peruse the app(s) you need to use (familiarity → confidence → proficiency).
5. Write down anything you are not sure about and ask for clarification.
6. Have a clear objective before you reach for a device or approach a computer.
7. Break down multipart objectives into manageable pieces — and sketch them out on paper — before reaching for a device or using a computer.
8. Know that applications are often flawed (*expect the unexpected*).
9. Get good help if you need it.
10. Document any technology-related kerfuffles and/or revelations into your *technology notebook*.

Troubleshooting

Let's say you are working calmly & methodically toward completing your task at hand when something stops you dead in your tracks; something **derails** your *Getting Stuff Done With Technology Train.*

Examples include:

- error messages
- "resource not found" messages
- system update messages
- *bloatware* update messages
- scam virus alert messages
- frozen device or app

Clarity and documentation are essential

“DETECTIVE TABBY”

If you have a smartphone (or anything that takes a picture), take a photo of the screen.

Write down (or take pictures of) any error messages that give you pause.

If a cryptic message has a close button (**X**), click or tap it and then, if possible, continue working.

If you are **really** unsure about something, then get *good help* (the sooner the better).

Depending on the severity of the issue, this could be in form of your organization's tech support and knowledgeable co-workers (and preferably **not** *know-it-alls* and *well-meaning jackasses*).

Proceeding mindfully is key to your success (remember that our devices and apps *only know what we tell them*).

THIS SHOWS THE PROPER MINDSET FOR PRODUCTIVE, STRESS-FREE TECHNOLOGY USE

Review your current objective and check your work before submitting it.

Case Study: Working From Home With a "Gifted" Computer

Even if you are not (from a technology perspective) a "blank slate" as my good friend Barbara is, please note that the beginning is not only a great place to start, it's a great place to **regroup** (with a nice cup of tea).

ILLUSTRATION BY DINORA NURTDINOVA

What to do if somebody gives you a computer or mobile device (or some other gadget)...

"HERE'S THE PLAN"

1. Get a *technology notebook.*
2. Get any administration-related *usernames* and *passwords* from the person who gave you the computer, tablet, or other gadget.
3. Make a nice cup of tea.
4. Use the technology basics (🐭 ☝️) to noodle around the computer or mobile device for the purposes of getting familiar with it and to make sure it is in good working order (e.g. not sluggish and no pop-up ads).

If the computer is sluggish and/or there are telltale signs of *bloatware* and/or *malware*, then *factory reset* the computer (which I would recommend doing anyway as a matter of course; clean slates and fresh starts can be extremely beneficial for devices as well as their human users).

After Barbara had recorded her *administrative username and password* (and *login info*) for her notebook computer into her technology notebook **and** her computer had been determined to be (or made to be) in good working order **and** free from *malware* & *bloatware*, Barbara was now ready to get to work.

Barbara's course of action

1. Make another nice cup of tea.
2. Document her job requirements (including which apps will be needed).
3. Ask her employer about *file sharing* and *file types* (e.g. .docx, .rtf).
4. Peruse her new computer's *file structure*
5. Noodle around with the apps that she needs to use.
6. Think about *workflow* (e.g. create a document → format the document → save the document → email the document).
7. Ask employer for clarifications.
8. Document all answers and any work-related insights gleaned along the way.
9. Barbara should also review her technology notebook on a daily basis.

Whenever you are exchanging files with somebody, ask them about their preferred *file types* and *fonts*.

It is always important to know what you are working with (e.g. *devices, apps, online services* and *file types*).

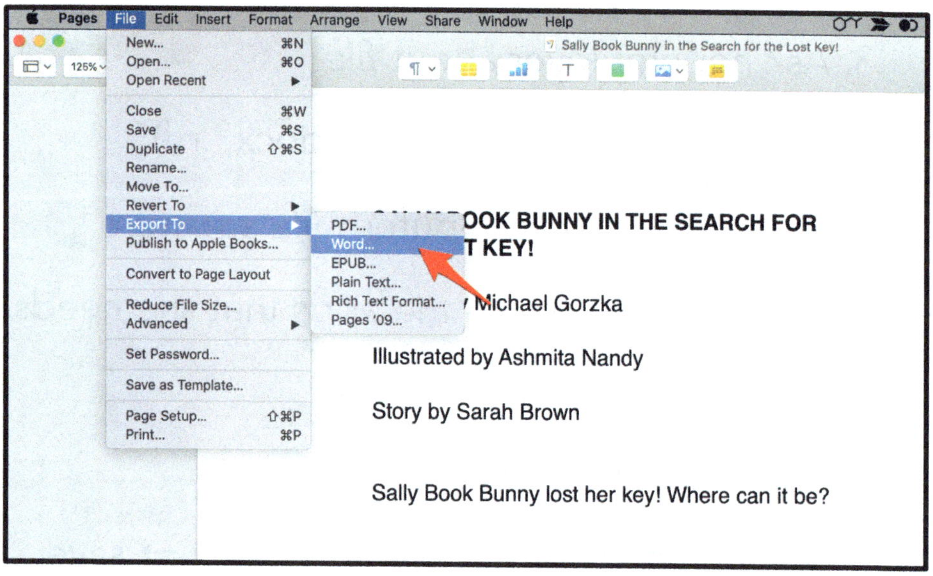

Common file types (+ the Mystery File)

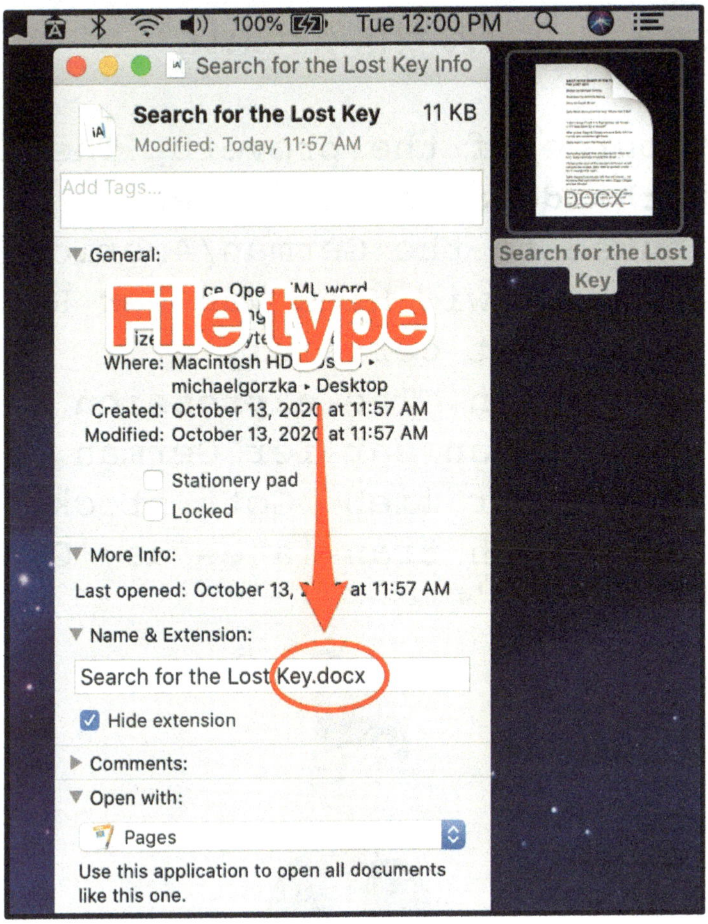

A CONTEXT MENU USED TO REVEAL A FILE'S FILE TYPE

"The devil is in the details"

"The source of the proverb 'The **devil is in the details**' is often attributed to the German/American architect Ludwig Mies Van Der Rohe. This is almost certainly a misattribution. The **expression** derives from an earlier German proverb – "Der liebe Gott steckt im **detail**", which translates as 'God is in the **detail**'.[12]

"BOW TIES GANG"

[12] Martin, Gary. "'The Devil Is in the Details' - the Meaning and Origin of This Phrase." *Phrasefinder*, www.phrases.org.uk/meanings/the-devil-is-in-the-details.html.

Despite our advanced gadgetry, nitty gritty details are so very important for productive, stress free technology use.

Many of my clients have given them short shrift and experienced considerable vexation and lost time as a result.

"NON RECYCLABLE TIME"

When the organization you work for gives you info you need for your job, you need to really lock it down so to speak. Record said info into your technology notebook → and take the time to review it and understand it.

For example, Barbara's company gave her an email address.

Barbara **could** access her email through a website with a username and password.

Barbara **could** access her email through a mail app on her computer through her computer's "Settings" or "Control Panel" or "System Preferences" (names are often arbitrary but functionality will always be highly relevant).

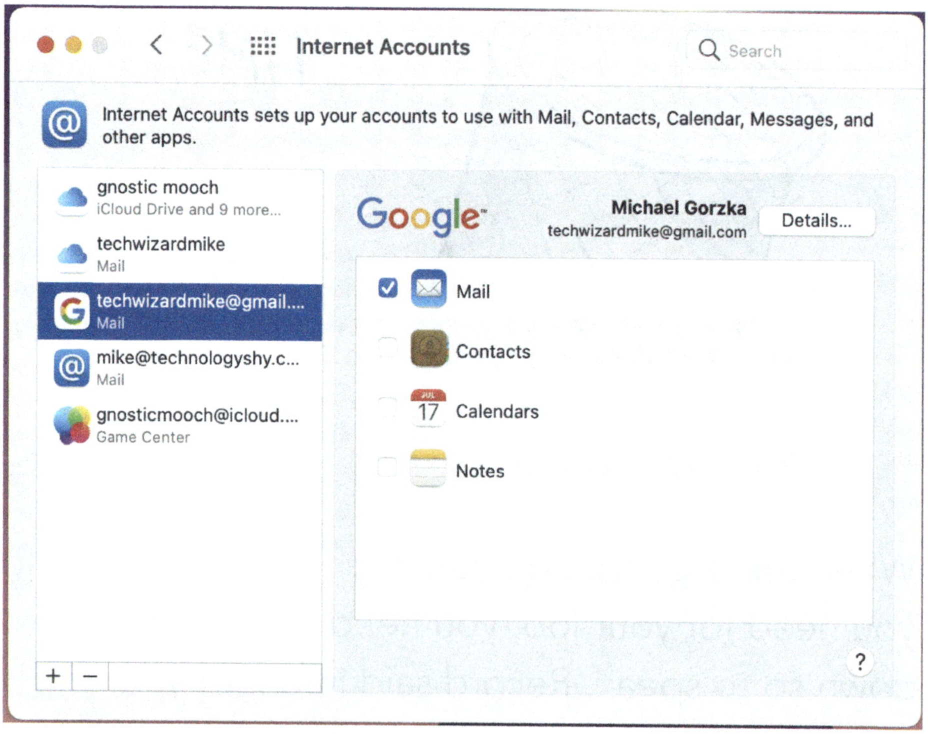

EMAIL SETTINGS ON TECH WIZARD MIKE'S COMPUTER

My goodness, if I had a nickel for every client who could not access their email due to forgotten passwords and/or lack of knowledge as to **how** they are accessing their email through apps on their computers or mobile devices…

… I'd have a lot of nickels!

Barbara's job requirements:

You must know what is expected of you and how you will use technology (in the form of computers, mobile devices, apps and websites) to fulfill those expectations.

- Email
- Access online resources (through websites)
- Create and format documents (e.g. docx file type, 12pt Times New Roman font)

Barbara carefully recorded her job duties into her technology notebook → along with any other pertinent information (e.g. web addresses, usernames and passwords).

The term "username" was unfamiliar to Barbara so she asked her Tech Wizard friend what it meant.

Barbara also needed to be able to create documents on her computer in a specific format (.docx), point (12pt) and font face (Times New Roman).

Barbara's new computer already had a word processing application.

It doesn't really matter which one (e.g. Microsoft Word or LibreOffice Writer).

What mattered is that Barbara could use it (after some *noodling around* with the word processing application's toolbar and menus) to create documents in the required specifications.

One of the keys to Barbara's success was her focus on *getting stuff done*.

Barbara made a nice cup of tea and casually clicked around the unfamiliar application's *interface* with zero expectations; simply taking in what she saw…

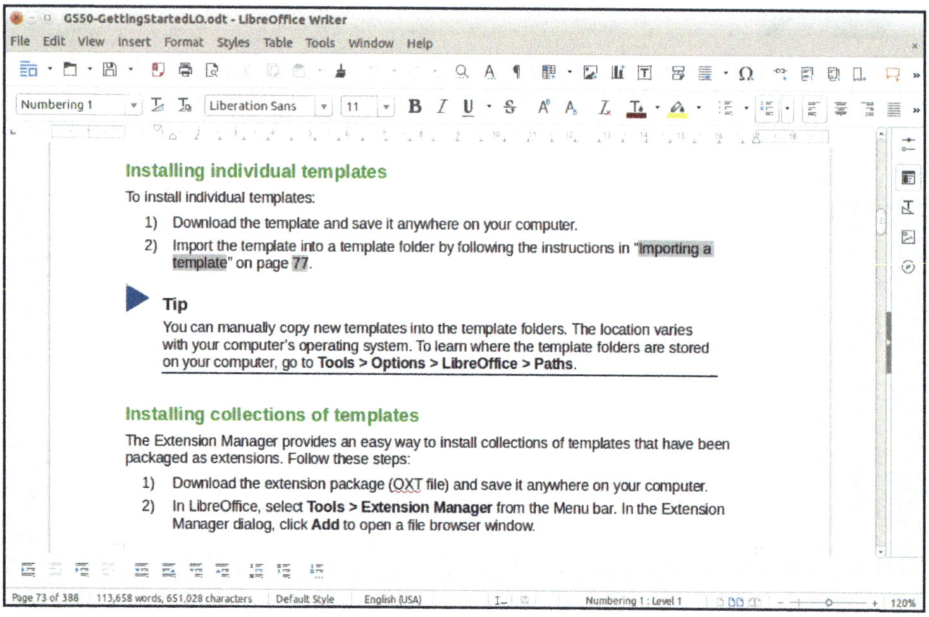

LIBREOFFICE WRITER

When she came across something that was really useful (e.g. a typeface box), she made note of it.

Barbara's word processing application had **a lot** of features (e.g. mathematical formulas, databases, and diagrams). Many of which she would never have any need for.

Barbara simply took what she needed and disregarded the rest.

Which OS is on your computer or mobile device?

CONTEMPORARY DEVICES SHOWING OPERATING SYSTEMS (WINDOWS, MACOS, IOS)

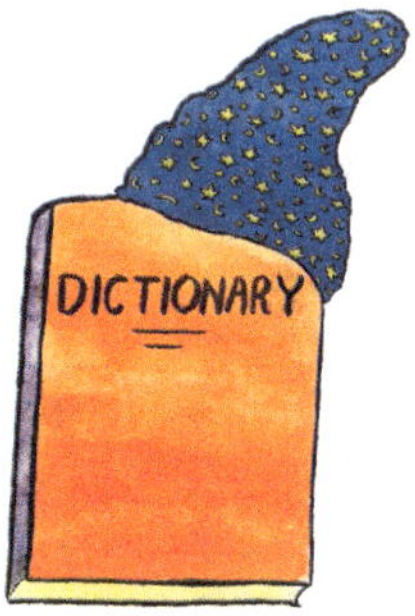

Operating System
Every desktop computer, tablet, and smartphone includes an operating system that provides basic functionality for the device.

Be very patient, methodical and visual during technology use

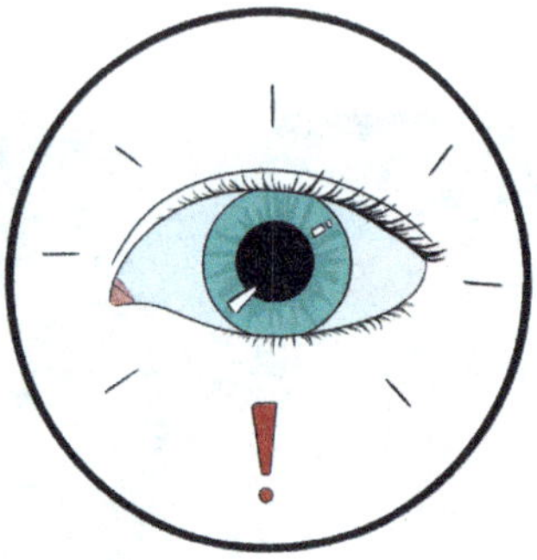

Regardless of which OS your device has, there will always be things that you can use to accomplish your current task at hand including...

- icons
- buttons (including *navigational breadcrumbs*)
- menus (remember that you have to *summon a context menu* 👩‍🍳)
- text fields (including *search boxes* 🔍)
- directions, clarifications and error messages

During the course of your technology travels (and occasional tribulations) *visual cues* will reveal navigation, functionality and yes, things to watch out for such as evil email attachments, phony-baloney "virus found" scams and other forms of technology-related shenanigans and skullduggery.

Barbara's Peripherals

You may not be able to use some *peripherals* (e.g. a mouse) with some devices (e.g. a notebook computer) due to a lack of *compatibility*. Although in some cases an adapter can be used to "bridge the gap" so to speak.

For example, Barbara could not get the hang of her notebook computer's trackpad and preferred the desktop mouse she had used in the past.

Compatibility issue

The mouse could not be *directly attached* to Barbara's notebook computer and so an adapter had to be procured:

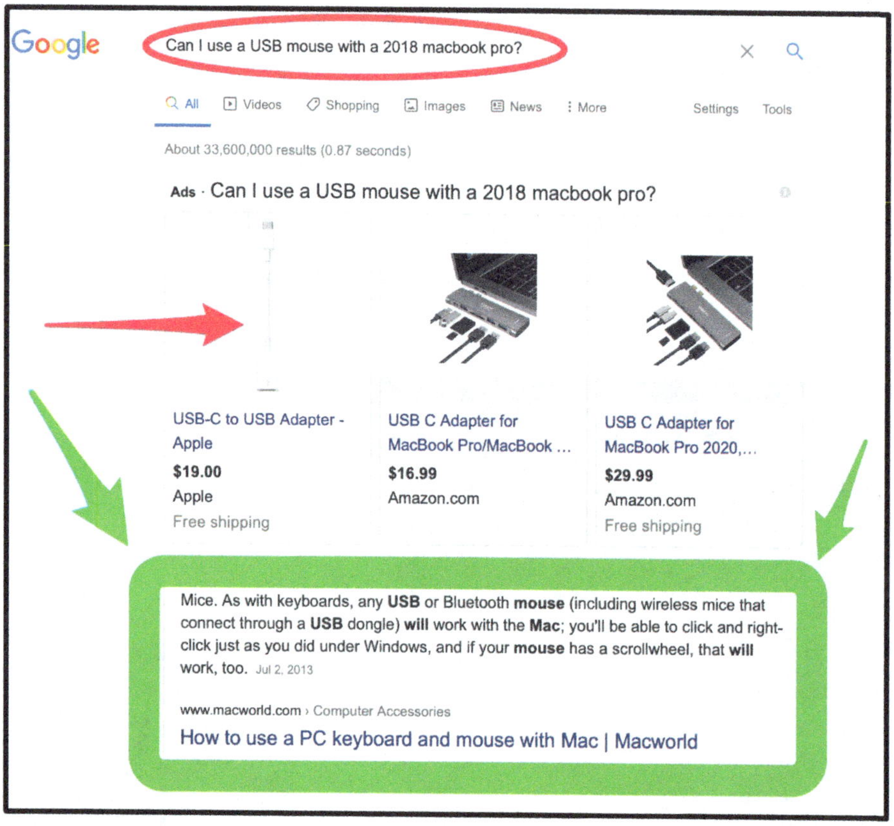

1. Know what you are using. 2. Know what you want to do. 3. And then search the web.

Should Barbara buy a printer?

All the documents Barbara would create need to be sent via **email attachments** (as per her employer's instructions which Barbara had diligently recorded into her *technology notebook*) but Barbara found it easier to proofread printed documents.

Once again *compatibility* came into play here as the printer could not be **directly attached** to Barbara's notebook.

(And even Barbara's Tech Wizard Tour Guide found the *wireless connectivity* to be arcane and very buggy).

Things to consider...

Printers are cheap but ink cartridges are très expensive 💰

Barbara also thought the ink cartridges would be difficult and tiresome to replace (after we watched an online video from the printer manufacturer on that procedure).

Workflow Solution*:*

Barbara felt that she would only need to print lengthy (e.g. five pages or longer) documents.

She could save those documents to a flash drive and then print them at her local adult community center for ten cents a page.

Beware of *gadget lust*. We should always think carefully before purchasing things we do not have a compelling use for.

Things to keep in mind...

- Some brands of devices (especially the lower cost ones) ship with *bloatware.*
- A storage device may inexplicably turn into a pumpkin.
- A storage device may contain software (e.g. automatic backup) that will not work properly
- When sharing documents with somebody, make sure your document is in their preferred *font* and *font size.*
- When you are using a work device (that is a device that was assigned to you by an employer), **work** is all you should use it for ❣
- Use your work email address for work only; and keep your messages professional ❣
- Always proceed thoughtfully and methodically during technology use.

Technology Project Case Study

"CUT THE CABLE"

More specifically (and it is great if not essential to be **specific**), our example objective is to watch the Turner Classic Movies station on TV without paying for cable or satellite.

Taking stock of what we currently have that could be useful...

- a MacBook Pro (running macOS Catalina)
- a high-speed (at least it should be) Internet connection
- an HDTV

What may be needed for any project...

- additional hardware
- a specialized app
- an online subscription service
- correct terminology

You may need to translate what you would like to do into specific technology terminology — which in this example is "live streaming" 🤓

“WHAT DO YOU CALL IT WHEN YOU WATCH A TV CHANNEL ON YOUR COMPUTER?”

Start from the Drawing board

Conceptualize 🙌 and then write down (or sketch out) what you would like to do 🖍️

"PROJECT DRAFTING"

So that you can begin your research by asking a focused question.

For example:

How can I stream Turner Classic Movies?

A possibility...

A web search and some well-written and objective articles on the subject lead us to several online services with this one being the most promising...

This is not an endorsement, by the way, we are only using "sling" as an example.

(Indeed "sling" could be no more by the time you read this or it could have new management or it could have adopted a new business model, and the list goes on.)

Stop and evaluate, refine your objective, and if necessary, do more research

This service will enable us to watch Turner Classic Movies on our computer 💻 → but we want to be able to watch it on our TV 📺

How to stream a MacBook Pro to a TV?

RESEARCHED POTENTIAL ANSWER: AN APPLE TV

Now... As always, proceeding *methodically*, before we purchase any new hardware or services, we need to find out: Is our TV *compatible* with the Apple TV?

plug and play

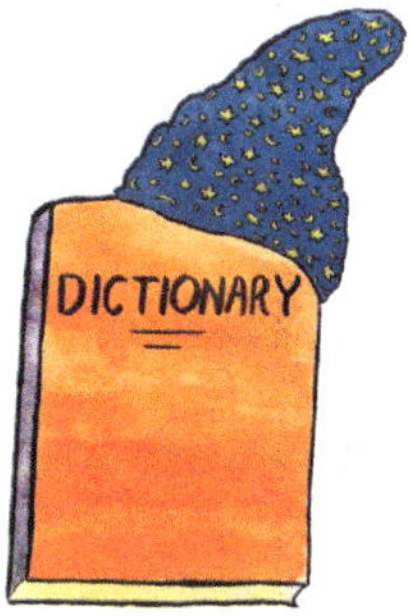

denoting or relating to software or devices that are intended to work perfectly when first used or connected, without reconfiguration or adjustment by the user.

Hardware evaluation 👀

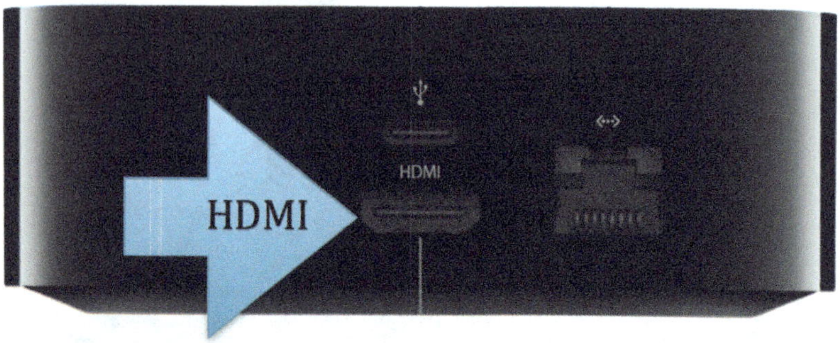

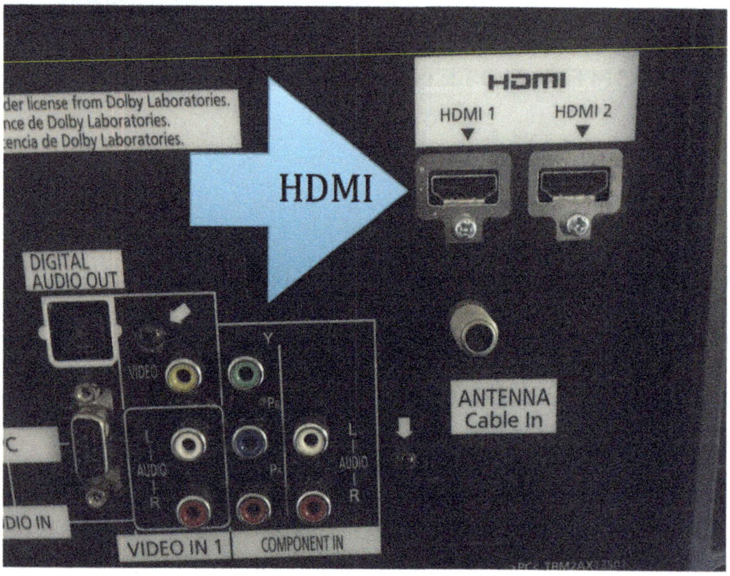

We have matching HDMI ports on both the Apple TV and the TV itself❗

This being the case, the TV may automatically "recognize" the Apple TV (which is a *plug & play* device) once the two are connected.

Things to remember for any online service consideration:

- Remember to calculate total costs
- Carefully read the *fine print* and understand what you are getting.
- Record the particulars of any service in your technology notebook (e.g. name, what it does for you, web address, trial expiration date).
- After signing up for any online service, periodically review your bank / credit card transactions (which we all should be doing anyway)
- Evaluate the service and then make an informed decision if you want to continue using it.
- Remember to get *good help* if you are unsure about anything.

Other factors may come into play...

In our example, the service might be great but our Internet service might be crummy and/or a no-goodnik might be stealing our WiFi.

"THE WIFI THIEF"

What you need to know...

- Online services can (and will) frequently change.
- Some online services may not be reputable.
- Business mandates might come into play (e.g. a particular service may not be available for a particular device or for an operating system or for a physical locale).
- Hardware can become *obsolete* and/or no longer supported by its manufacturer.
- An *OS update* could cause an app on your device to longer function properly (if at all).
- Expect the unexpected 😯 (e.g. websites being down; internet-hogging neighbors; sudden price hike; discontinuation of a service or a particular aspect of a service that you really liked).

"Simplicity boils down to two steps: identify the essential, eliminate the rest." — Leo Babauta

“WRAP IT”

1. An idea 💡 leads to a visualized outcome 🙌

2. Find out the apropos tech terms 🗣️ and then research 🔎

3. Write down or sketch out what you need to realize your current objective.

4. Research and (if necessary) purchase any additional hardware and services → and record them into your *technology notebook*.

5. Make a cup of tea and proceed methodically through any installation procedures and/or physical setup.

6. Carefully follow the visual cues on any app(s) or online service(s) that you need to use to complete your objective.

7. Seek out *good help* when you feel you need it.

Another Note on Pedestrian Crosswalks

As you are waiting to cross a busy intersection, you might idly wonder about the spectacular yellow 1972 Volkswagen Super Beetle that just majestically drove by…

"VW BEETLE APPEARANCE"

But the unexpected sight of that marvelous vehicle would not completely take your mind away from your **primary objective** — which is crossing the street.

There are times to noodle around 🍜 And there are times to get stuff done ✔

And when you need to **get something done** — such as attach a photo to a text message — you should not allow yourself to be needlessly distracted by things on a screen that have nothing to do with the task at hand.

Granted, you will need to peruse any unfamiliar screens (or *interfaces*) that you land on.

And you may wonder why some of the things are there (some of which may not even be functional or have any discernible purpose 🤨).

You can record anything that distracts you into your technology notebook for later review & clarification —> getting it out of your head and into an organizational system that you can trust.

Technology success = defined objective **+** technology notebook **+** focus **+** proceeding methodically, following *visual cues*:

"TUFFY ON THE TRAIL"

Also be very observant if not outright vigilant during technology use; be mindful of things like a *new web browser tab* opening in a web browser window after you click a link or button.

← "So many buttons" 😟

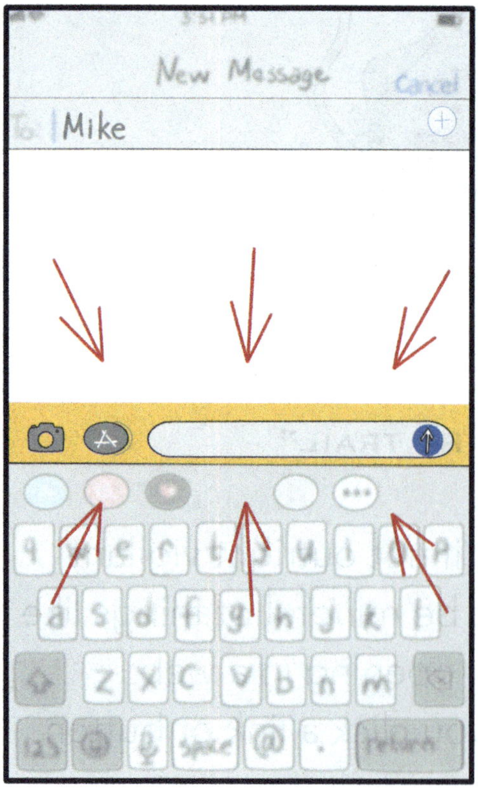

← The power of focus 🎯

On the Subject of Cloud Computing...

Did you know that some people do not own a computer, a tablet, or even a smartphone?

And yet, they are still able to *get stuff done* here in the twenty-first century.

This happens through *cloud computing.*

```
In the simplest terms, cloud
computing means storing and accessing
data, files and applications over the
internet instead of on the computer
or mobile device.
```

Nothing is actually stored on the computer when you use a cloud service that offers both applications (e.g. word processing, spreadsheets & presentations) and file storage.

The above screen capture shows a document open within a word processing application that is itself open within a web browser window.

Cloud users can login into their accounts and access their applications and files through any internet connected computer or mobile device via a web browser.

Cloud computing procedure

You can either buy an *economy computer* or find out if your local library and/or adult community center offers public computers that are well-maintained.

You can tell if a computer is well-maintained by its responsiveness and the absence of pop-up ads and warning messages.

You also want to feel comfortable as frankly public computer spaces can be hangouts for hooligans.

You could also bring your newly-purchased (and perhaps *factory reset*) economy notebook computer to your library or any safe place that offers free wi-fi.

ILLUSTRATION BY MARIA FEDOTOVA

Visit www.technologyshy.com/cloud for your Tech Wizard Tour Guide's cloud computing how-to videos 🍿

Create a cloud services account and write down your username, password and the web address for the cloud service.

After your cloud services account has been established, getting stuff done with cloud applications is really just a matter of…

1. having your objective firmly in mind or written down (e.g creating a presentation).
2. logging into the public computer (if applicable)
3. starting the computer's web browser & navigating to the cloud services web site
4. logging in with your username and password
5. following visual cues within the cloud application to work methodically toward completing your objective
6. *backing up your files* to an external storage device (recommended)

Small Screen Usability Issues

"WTF?"

Many of my technology shy friends are using small screen devices (e.g. tablets and smartphones) for online services / websites that really need a larger screen (e.g. a desktop or notebook computer).

And the primary reason these online services / websites need a larger screen is because they are poorly designed (there I've said it).

And even with a larger screen, these online services / websites can be challenging to use 🤕

Want a plan? Here's a plan...

Buy an *economical notebook computer*. They are much easier to use for many online resources and services.

If you buy a notebook computer that is running a Windows operating system, *factory reset it* → as previously-owned computers could be infected with malware and/or bloatware and new computers could have bloatware pre-installed in the form of "trial versions" 😒

Where to Go From Here

Go to www.technologyshy.com for…

* Help for the Technology Shy: A commonsense approach to safe, productive, stress-free technology use
* Tabby's Guide to Thoughtful Tea Drinking
* Technology Basics DVD 
* Technology tips & tutorials

Visit your local library or adult community center if you would like some help navigating the World Wide Web (i.e. visiting websites).

Help for the Technology Shy

A commonsense approach to safe, productive, stress-free technology use

Easy (and fun 😀) to read, 436 pages, indexed, large print, lots of colorful illustrations.

In this book, you will find some baby-boomer flashbacks, a two-headed alien, at least two dinosaurs, instructions for brewing the perfect cup of tea, a discourse on rocket science 🚀 (not really 🙂) and an emotional reset button — but you will also get some rock solid technology help and guidance (applicable to all devices 😯) using real life case studies and examples.

"DINOSAUR IN THE BUILDING!"

Technology Basics DVD

In this one-hour Help for the Technology Shy program, we describe and demonstrate how to make thoughtful and productive use of...

- the computer mouse
- trackpads
- touchscreens (or touch screens)
- keyboards

This instructional DVD is designed for technology novices. But this technology help program can also help anyone who is struggling with contemporary technologies.

Tabby's Guide to Thoughtful Tea Drinking

A PASSWORDS BOOK CLEVERLY DISGUISED AS A TEA DRINKING BOOK 🤫

Yes, in this book you will have 100 immaculate blank pages for your logins, passwords, answers to security questions, web addresses and whatever else you need to keep track of → but the first 40 pages provide a primer for safe, productive, stress-free technology use.

Check it out:

www.technologyshy.com/tabby

More(!) Help for the Technology Shy

“BLOGGER”

Tips, tutorials and how-to videos which supplement this book are on Tech Wizard Mike’s blog.

Italicized words and phrases in this book are blog & video topics.

Check it out:

www.technologyshy.com/blog

Thank you for reading this book! We sincerely hope you found it to be helpful.

Say goodbye to drama, toxic people, and self-criticism. Say Yes to more happiness, love in your life, and time with good friends.

“GOODBYE, DRAMA”

Index

"A FLIGHT OVER THE BIG WORLD"

Made in the USA
Monee, IL
02 November 2026

67383861R00116